TELEVISION
AND THE EARTH

TELEVISION
AND THE EARTH

Not a Love Story

Jennifer Ellen Good

Fernwood Publishing
Halifax & Winnipeg

Editing: Brenda Conroy
Cover design: All Caps Design
Printed and bound in Canada by Hignell Book Printing

Published in Canada by Fernwood Publishing
32 Oceanvista Lane, Black Point, Nova Scotia, B0J 1B0
and 748 Broadway Avenue, Winnipeg, Manitoba, R3G 0X3
www.fernwoodpublishing.ca

Fernwood Publishing Company Limited gratefully acknowledges the financial support
of the Government of Canada through the Canada Book Fund and the Canada Council
for the Arts, the Nova Scotia Department of Communities, Culture and Heritage,
the Manitoba Department of Culture, Heritage and Tourism under the Manitoba Book
Publishers Marketing Assistance Program and the Province of Manitoba, through the
Book Publishing Tax Credit, for our publishing program.

Library and Archives Canada Cataloguing in Publication

Good, Jennifer Ellen, 1965-
Television and the earth: not a love story / Jennifer Ellen Good.

Includes index.
ISBN 978-1-55266-552-7

1. Television--Social aspects. 2. Materialism--Social aspects.
3. Environmentalism on television. 4. Human ecology on television.
I. Title.

PN1992.6.G66 2013   302.23'45   C2012-908132-9

CONTENTS

For my daughter and all young people:
may your relationship with the Earth be a love story.

ACKNOWLEDGEMENTS

I have pondered the ideas in this book for years; over those years many people have inspired and supported me. Professors, advisors, students and colleagues at the University of British Columbia, York University, Cornell University, Royal Roads University and Brock University have shared their wisdom and helped me hone my skills. Co-workers at Pollution Probe and Greenpeace were exemplary role models. My family and friends have been patient, and challenging, as I endlessly critiqued and analyzed. I thank my parents for their example of questioning everything and seeking justice everywhere (and for reading the early stages of the manuscript!). I also thank the fine folks at Fernwood Publishing for helping me turn a manuscript into a book. And to Ana I offer the clichéd sentiment, and fundamental truth, that the sharing of these ideas would have been impossible without her.

1. WARNING

In 1994, a group of scientists, including 104 Nobel laureates, offered a "Warning to Humanity" that "human beings and the natural world are on a collision course" (Union of Concerned Scientists 1994). Eleven years later, 1,300 scientists and social scientists — the largest group of scientists ever assembled to assess human impact on the environment — concluded: "Human activity is putting such a strain on the natural function of Earth that the ability of the planet's ecosystems to sustain future generations can no longer be taken for granted" (Millennium Ecosystem Assessment 2005). In March 2011, a dozen scientists writing in the journal *Nature* wondered: "Has the earth's 6th mass extinction already arrived?" They answered their question by noting that indicators "suggest that a sixth mass extinction may be under way, given the known species losses over the past few centuries and millennia." If scientific warnings aren't enough, there's Ronald Wright's admonition in *The Short History of Progress*: "The most compelling reason for reforming our system is that the system is in no one's interest. It is a suicide machine" (2004: 131).

Our mass-extinction, collision-course suicide machine hasn't always been our mode of travel. Nor has destruction always been our destination. In fact, humans lived for millennia in a mutually respectful co-existence with the planet. For example, "culturally modified trees" are found in many parts of the world, including the forests of British Columbia. There I was shown huge cedars with long dark swaths where strips of bark were removed hundreds of years ago — to make clothing, containers, adornments. And the trees thrive to this day. This was the idea behind the practice: make use of the trees without destroying the forests. Could there be a practice more antithetical to our current forest activities? The term "clearcut" was coined to describe what we do to huge forests. In fact, Greenpeace estimates that every two seconds an area of forest the size of a football field is taken down due to logging or other "destructive practices" (Greenpeace n.d.). According to the 2010 Global Forest Resources Assessment, "the world's forest biodiversity is threatened by a high global rate of deforestation and forest degradation as well as a decline in primary forest area" — resulting in the loss of approximately two-thirds of the world's original forests (Food and Agriculture Organization of the United Nations 2010).

It is not only in the forests where we can question the sanity of current practices — or clearly see our collision course with nature. While our seemingly endless oceans were once filled with shimmering fish, the United Nations

1

Food and Agriculture Organization states that 96 percent of the world's fish stocks are in trouble (either "moderately-exploited," "fully-exploited," "over-exploited" or "depleted"). In their *New York Times* article "Has the Sea Given Up Its Bounty?" William Broad and Andrew Revkin highlight that, not unlike the clearcutting of forests, with

> precise sonar and navigation gear, more than 23,000 fishing vessels of over 100 tons and several million small ones are scouring the sea with trawls that sweep up bottom fish and shrimp; setting miles of lines and hooks baited for tuna, swordfish and other big predators; and deploying other gear in a hunt for seafood in ever deeper, more distant waters.

(And in a perversity that only makes sense in this "modern age," in spite of the depleted fish stocks around the world, I can walk across the street this very moment and buy a Filet-O-Fish from McDonald's for $2.99, or just a little more for a *Double* Filet-O-Fish.) Finally, in some kind of Hollywood-esque irony, perhaps the best example of modern civilization as a suicide machine is the fact that the very foundation of our society — the burning of fossil fuels — is changing the climate of the entire globe.

But a detailing of our environmental status isn't what I want to explore in this book — not exactly. My guess is that you have already heard a lot about environmental degradation. It's hard to avoid talk of climate change and habitat loss, hard to avoid the appeals to recycle and bring your own bags. No, what I am interested in exploring isn't so much "where we're at" but this question: *How did we get here*?! The essence of the answer I propose is contained in one word: stories. Specifically, the answer is television's stories and the role they have played in making us believe in materialism. It is materialism that has devastated the planet. Making a case for why television's stories are responsible for our materialism, and why our materialism has caused our current reality of travelling in a mass-extinction, collision-course suicide machine, is what this book is about.

A handful of other books have appeared over the years offering cautionary tales about television, including Jerry Mander's *Four Arguments for the Elimination of Television* (1977), Marie Winn's *The Plug-in Drug* (1977), Neil Postman's *Amusing Ourselves to Death* (1985), Joyce Nelson's *The Perfect Machine: Television in the Nuclear Age* (1987), Bill McKibben's *The Age of Missing Information* (1993) and Richard Zurawski's *Media Mediocrity* (2011). These books have often been ignored by the academic community and, in the case of the more dramatic television critiques such as Mander's call for the elimination of television, somewhat derided by the "serious" television research community. (One of my PhD committee members told me that he would leave my committee if Jerry Mander appeared anywhere in my dissertation.) The problem with this situation is that the "serious" television research published by the academic community tends not to be very accessible to the vast majority of people. On the other hand, the more popularly available television critiques have been few and far between and, as I point out, have often been overlooked by the academic community.

Television and the Earth, therefore, creates connections. It connects those who may not spend a lot of time digging through academic books and journals with academic television research. It also connects television research with an underexplored outcome of television viewing: materialism. Finally, this book connects materialism and our dire environmental situation.

The clichéd advertising pitch encourages us to buy something because the product is "new and improved." The three thematic areas explored here — television, materialism and the environment — are certainly not new, and, indeed, they encompass vast areas of research and writing. And I am not claiming to be an expert, or to offer "new and improved" ideas in all of these areas. I am, however, reworking and connecting themes, research and ideas in each of these areas. In this respect what I offer is, indeed, new and improved.

Television and the Earth includes four more chapters after this one: the Television chapter, the Materialism chapter, the Environment chapter and the concluding chapter. In the Television chapter (Chapter 2), I discuss how television adversely affects us. I create a context for television's role as a storyteller in our lives, look at how television research is done, provide an overview of some of the theories of television's effects and offer research results that show a relationship between television viewing and our thoughts and behaviours.

One of the most interesting aspects of research into the effects of television consumption is just how difficult it is to do. The essence of an experimental design to study effects (i.e., research that tries to answer whether one thing — perhaps a toxin, a parenting practice, a form of mediated communication — is affecting another thing — perhaps someone's health) is the following: one group, the "experimental" group, is exposed to what you are testing; the other group, the "control" group, is not exposed to what you are testing. Then you look for any differences between the two groups. The challenge is that there are not many people, especially in a more affluent part of the world like North America, who do not watch television. In fact, as I highlight in the Television chapter, Nielsen estimates that as of 2010, close to one hundred percent of American homes had at least one television set (of those, 28 percent had two television sets and 55 percent had three or more television sets) (2010b). Nielsen's research also indicates that these television sets are being well used. In 2010 the average person in the United States watched 143 hours of television each month (2009a) and the situation shows no signs of changing. In 2012 Nielsen reported: "tv holds the lion's share of ad dollars and consumers' media time." Thus, the creation of a "control group" made up of individuals who do not watch television is virtually impossible.

Of course this is not to say that television research is impossible. Far from it. There are all kinds of ways to explore the relationship between television and its viewers, and a tremendous amount of effects research has been done in television's relatively short life. In fact, television is probably the most studied communication medium and for good reason; as the statistics highlight, television plays a critical role in our lives.

The Materialism chapter (Chapter 3) revolves around an exploration of our relationship with the material world. How have we come to understand and make sense of the stuff with which we co-exist? As Annie Leonard points out in her popular *Story of Stuff* video, "We have become a nation of consumers. Our primary identity has become that of consumer, not mothers, teachers, farmers, but consumers. The primary way that our value is measured and demonstrated is by how much we consume. And do we!" Indeed, I believe that the shaping of the relationship between us and our stuff has been television's most powerful role, and yet, interestingly, there is a dearth of research on the topic. In this chapter, therefore, I make a case for the vital importance of understanding this role.

In the Environment chapter (Chapter 4) I discuss what I see as the most significant implication for television's story of materialism: how materialism affects our relationship with the natural environment. I believe that the environmental crisis is, fundamentally, a crisis of over-consumption, and nothing teaches us about why and how we should consume more clearly and compellingly than television's stories. In the concluding chapter (Chapter 5) I explore where we can go from here.

The reworking and reconnecting of established ideas can be challenging. Any challenge to entrenched ways of thinking about and doing things is bound to be provocative; even seemingly benign ideas that question the status quo can get a fierce reaction. For example, a few years ago I wrote an open letter to the Canadian grocery tycoon Galen Weston that appeared in the June 29, 2008, online version of the *Globe and Mail*. The letter proposed that the next big idea in ecologically conscious grocery stores should be refillable containers (i.e., moving up a step in the reduce, reuse, recycle hierarchy from recycling food containers to reusing them). The short letter, which seemed to me to be a fairly innocuous offering of a common-sense suggestion, was quickly greeted with almost a hundred responses, many of which were hostile and included personal attacks. That experience served to reinforce what I already knew: people can be very attached to the status quo, even when by certain objective measures the status quo is not serving them well.

If a seemingly benign suggestion that grocery stores play a more active role in reducing waste by facilitating refillable containers resulted in a hearty, and not always pleasant, outpouring of opinion, then the content of the next few chapters should engender some passionate responses. The propositions contained in this book are not even remotely benign or innocuous, but rather ask us to take a fresh look at, and ask fundamental questions about, our lives, what we value and the state of the planet on which we live.

So, how *are* television's stories destroying the Earth? This is the question I explore in the rest of the book.

2. TELEVISION

In 1977, when Jerry Mander published *Four Arguments for the Elimination of Television*, his task was Sisyphean. Back then television was fresh, new and somewhat magical; families were coming together around the glowing screen and loving the experience. The fact that Mander's arguments were often hyperbolic and sensational didn't help. For example, his third argument for the elimination of television commenced with a chapter entitled "Anecdotal Reports: Sick, Crazy, Mesmerized." Not surprisingly Mander's sensational case for getting rid of television altogether did not hold a lot of sway with the many people who were loving their televisions, or with those who were serious about their television research.

In some ways, my task in this chapter should be much easier than Mander's. I too want to make the case that we are being adversely affected by watching television, but I have two very important realities working in my favour. First, you probably already have a sense that television viewing can be problematic. Second, there is credible research to back up that sense. Indeed, some reputable researchers now even suggest that, under certain circumstances, it is wise to stop watching television completely. For example, both the American Academy of Pediatrics and the Canadian Pediatric Society offer strong caution to parents regarding television viewing — with the American Academy recommending that parents "discourage television viewing for children younger than two years" (2001: 424).

One of the concerns that researchers have highlighted for television viewers, young and old, is the way in which the medium encourages a sedentary lifestyle — and our sedentary lives have been linked to serious health concerns, such as epidemic rates of obesity and type-2 diabetes. In their seminal study published in the journal *Pediatrics*, William Dietz and Steven Gortmaker found that "in 12 to 15 year old adolescents the prevalence of obesity increased by 2% for each additional hour of television viewed [per day]" (1985: 807). A similarly daunting indictment of television's health impact on adults can be found in the meta-analysis undertaken by Anders Grontved and Frank Hu in 2011 and published in the *Journal of the American Medical Association*. The study involved grouping together previous research in order to explore the associations between television viewing and risk of type-2 diabetes, cardiovascular disease and all other causes of death. The results confirmed that, indeed, "longer duration of TV viewing time is consistently associated with higher risk of type-2 diabetes, fatal or nonfatal cardiovascular disease, and all-cause mortality" (2011: 2454). To put this in perspective: in a town of 100,000 people, as average television viewing increases by two hours per day, type-2 diabetes increases by 176 individuals, fatal cardiovascular disease by 38 individuals, and 104 additional individuals die prematurely from all other causes.

To these medical concerns, one can add the multitude of studies that have explored the relationship between television and violence (arguably the most studied aspect of television). These studies have clearly shown that television viewing is related not only to changes in attitudes about aggression and violence, but also to increases in aggressive and violent behaviours. In fact, George Gerbner and colleagues coined the term "mean world syndrome" to describe their consistent findings that when heavy and light viewers of television are compared, heavier viewers think the world is a "meaner and scarier" place (Gerbner, Gross, Morgan and Signorielli 1980). Longitudinal studies by Rowell Huessman and colleagues have allowed for the "unequivocal" conclusion that "violent television... increases the likelihood of aggressive and violent behavior in both immediate and long-term contexts" (Anderson, Berkowitz et al. 2003: 81). As well, research links television viewing to issues of concern as diverse as poor learning and academic performance; racism and sexism; tobacco, drug and alcohol use; poor personal body image and eating disorders; and materialism (see the American Academy of Pediatrics 2001 policy statement "Children, Adolescents and Television" for an overview of television effects research on young people).

Why Care about Television Anymore?

In the age of Facebook, Twitter, YouTube, and texting (to name but a few), you may be harbouring a fundamental question as you read about television: Why, in the era of digital communication technology, should we be talking about such an "antiquated technology" as television? The answer is that television is, arguably, not only a more dominant force in our lives than ever before but *the* most dominant communication medium in our lives. In the second quarter of 2010, the U.S.-based television research firm Nielsen found that the average American consumed 143 hours of "in-home" television each month. This is approximately 35.75 hours a week or just over five hours of television viewing a day. This is up significantly from the four hours of television watching per day that Jerry Mander was concerned about in the mid-1970s. Nielsen found that in 2008 Americans were watching more television than at any other point in history (Nielsen 2009a), and in 2012 Nielsen concluded that television continued to consume the "lion's share" of viewer attention and that advertisers were responding by spending more on television advertising than all other advertising platforms *combined* (Nielsen 2012). Therefore, while new forms of communication technology have caught our attention and have become essential to our lives, television, in many ways, continues to reign supreme. People spend more time watching television each day, on average, than doing anything other than sleeping and being at school or work.

So, part of the answer to "why talk about television now?" is that we're watching it more than ever before. We are also watching it on more and bigger screens than ever before. Fifty-five percent of American homes have three or more television sets, 28 percent have two sets, and only 17 percent have one set. This means that the average American home now has just over 2.9 television sets, which is greater than the average of 2.5 occupants in a home (Nielsen 2010b). And according to DisplaySearch (2006), the average North American's television screen is 29.5 inches. North Americans also own over half of the world's 40 inch (and greater) television screens and over 80 percent of the world's 50 inch (and greater) television screens.

Another reason to talk about television now is that we are watching television in more ways than ever before. We can watch television content on a conventional television set in "real time," but we can also time shift and use the digital video recorder (DVR) or personal video recorder (PVR) to watch television at our convenience. In addition to tracking our average "in home" television viewing, Nielsen tracks our "time shifted" television viewing, which, in 2010, added just over nine hours of viewing time each month. As well, television programming can be accessed on other screens — computer, tablet, phone, PDA (personal digital assistant) — so we can watch television content pretty much anywhere and anytime. Even without our own portable screen, television reaches us in many locations outside of our home — airplanes, bars, taxis, waiting rooms and even bathrooms.

The varied ways we access television are highlighted by changes in viewing measurement technology. In the past, television research companies like Nielsen and the Bureau of Broadcast Measurement measured television viewing levels using a "people meter" — a little box that would sit in your home on top of your television set. When you came into the room to watch television you would push "your" button on the people meter and it would track what you watched and for how long. These days researchers make use of a "portable people meter" (PPM), which picks up a high pitched tone in the programming, inaudible to us, that tracks viewing wherever it happens to take place on any electronic medium.

In the twenty-first century we therefore continue to consume television "the old fashioned way" in record quantities, and, with the aid of new communication technology, we are accessing television content in many new ways as well. Far from being left behind by new communication technologies, these gadgets allow us to embrace television with unprecedented zeal and intimacy.

When television first appeared in our homes, it took a place of prominence in the living room. We huddled around the flickering screen and celebrated the shared event. As the years went on, television's bulky black box began to be relegated to rec rooms and basements. The arrival of sleek flat screen televisions has, however, brought a certain renaissance. Television sets have regained a place of honour in many living rooms, often above the fireplace mantle. They have also increasingly arrived in our bedrooms and our children's bedrooms. Victoria Rideout and Elizabeth Hamel found that 19 percent of children in the United States under the age of one had a television set in their bedroom (2006). In a 2008 *New York Times* article, Tara Parker-Pope noted: "By some estimates, half of American children have a television in their bedroom; one study of third graders put the number at 70 percent. And a growing body of research shows strong associations between TV in the bedroom and numerous health and educational problems." When not watching in our living rooms, bedrooms and kitchens (where the screen can be neatly nestled into our fridge), we can find our favourite television shows conveniently located in our purses, briefcases and backpacks.

In this chapter, therefore, I join those who have come before me in making the case that television's role in our lives should be taken much more seriously than it is. In fact, the research shows that we should be downright fearful of that role. The wealth of television research makes it easier now to explain how television adversely affects us than when Mander encouraged us to eliminate the medium thirty-five years ago; certainly, many more cautionary voices have joined Mander over the years. It is also clear, however, that our love affair with television is more firmly entrenched than ever.

The Power of Stories

How did we get here? Does it matter how we live while we are here? What happens when we die? Perhaps it is especially because of the latter question, and our universal ultimate demise, that our questions, and the stories we tell in response, are constant. Throughout human history we have gathered around the fire's glow and told our stories in an attempt to make sense of our life and our death. Communication scholar Walter Fisher calls us homo narrans, proposing that our storytelling is the essence of who we are: "All forms of human communication can be seen fundamentally as stories, as interpretations of the world occurring in time and shaped by history, culture and character" (1989: 57). Scientists tell stories of actions and reactions. Parents offer stories about how to be safe and thrive. Suitors try to impress with stories of success and achievement. Friends share stories of past adventures and future dreams. Ultimately stories are all we have and all we are.

Stories contain nothing more, and nothing less, than versions of reality. Fisher contends: "Human communication in all of its forms is imbued with mythos — ideas that cannot be verified or proved in any absolute way" (1985: 87). This is why there is great power inherent in who gets to tell their stories, whose stories are listened to and whose stories are acted upon. Plato may have been one of the first to note this power: "Those who tell the stories rule society." But many others have paid tribute to the power of the story. In a 1991 speech at Columbia University, the noted (and controversial) author Salmon Rushdie offered that: "Those who do not have power over the story that dominates their lives, power to retell it, rethink it, deconstruct it, joke about it, and change it as times change, truly are powerless, because they cannot think new thoughts." In his book *The Truth About Stories: A Native Narrative* (based on his CBC Massey lecture), Thomas King proposes: "The truth about stories is that's all we are" and he quotes Ashinabe author Gerald Vizenor: "You can't understand the world without telling a story. There isn't any center to the world but a story" (2003: 32).

No researcher is perhaps more associated with humans and their stories than Joseph Campbell. Throughout his life Campbell explored the role and content of stories in people's lives around the world. The commonalities of those stories, especially the common traits of the "hero," became his book *The Hero with a Thousand Faces* (1949). When film director George Lucas drew upon this book to craft his *Star Wars* stories, Campbell found himself, and his ideas, launched well beyond academia and into the world of Hollywood.

In *The Masks of God, Vol. 3: Occidental Mythology* (1964), Campbell articulated four functions of myth. Myths can be metaphysical (they awaken awe, especially through participation in myth-related rituals); myths can be cosmological (they allow for sharing what is known of the physical world); myths can be sociological (they validate and support the social order); and myths can be pedagogical (they teach and guide us through the various stages of life).

In *The Power of Myth*, a six-part documentary about Campbell's research, he is asked: "So we tell stories to try to come to terms with the world, to harmonize our lives with reality?" Campbell responds that we do, and then clarifies:

> People say that we're all seeking a meaning for life. I don't think that's what we're really seeking. I think that what we're really seeking is an experience of being alive, so that our life experiences on the purely physical plane will have resonances within our innermost being and reality, so that we actually experience the rapture of being alive.

Campbell, therefore, believes that our myths, or stories, can play a role in helping us to experience *the rapture of being alive*. I find this striking because it seems to me that while stories continue to be told for many reasons, stories that help us to experience the rapture of being alive are few and far between. In what follows I highlight three other aspects of our current stories that I believe are important: first, all stories are fiction; second, we hide our stories behind a veil of frivolity; and third, mediated communication allows for a very few storytellers to tell huge numbers of people the same stories.

By the fundamental fiction of all stories, I mean that all stories are crafted by us to share a particular point of view. Even when stories are seemingly about "hard fast rules," such as in the world of science, individuals still tell those stories from their own viewpoint. This is not to deny that there is a physical world, or in Campbell's words, a "physical plane," but even within that realm, the governing "laws" are, it turns out, much less linear, predictable and law-like than we once thought (and perhaps hoped for). Indeed, science stories are changing and shifting to deal with new storylines all the time.

Research in the area of quantum physics highlights that in our very attempts to tell cause-and-effect stories about the inanimate world, we affect what we are trying to know and measure. The "double slit experiment" illustrates this, showing that photons, one of the most fundamental types of matter out of which everything is made, behave one way when observed and another way when not observed. David Darling offers in the *Encyclopedia of Science*:

> If we don't ask where the photon is, it behaves like a wave; if we insist upon knowing, it behaves like a particle. In classical physics such a situation would be unthinkable, outrageous. Yet there it is: the act of observing light makes its wave nature instantly collapse and its particle aspect become manifest at a specific point in space and time. *It's almost as if a photon knows when it's being watched and alters its behavior accordingly.* (Darling n.d.)

The last part is worth emphasizing: It's almost as if a photon knows it's being watched and alters its behavior. This is hardly the linear cause-and-effect story scientists had been telling each other, and us, about the way the world works. The story of photons, therefore, needed to be revised.

My second point about stories is that we downplay and trivialize their role and their importance in our modern world. As the laugh track cues us when to laugh, the reality TV contestants vie for prizes by undertaking ridiculous stunts and the sound bite tells us the news, our stories seem silly. In his book *Amusing Ourselves to Death*, Neil Postman proposes that we have trivialized our stories to the point of turning them all into entertainment. What this means, according to Postman, is that "short and simple messages [have become] preferable to long and complex ones; that drama is to be preferred over exposition; that being sold solutions is better than being confronted with questions about problems" (1986: 131).

So while the stories we once told each other spoke to the essence of who we were, and our very survival, the vast majority of our stories are now created with one objective in mind: mass entertainment. Our stories have, therefore, been given a veil of frivolity. We have come to undervalue the so-cietal role of stories because any social function of storytelling has *seemingly* been subsumed by the grander scheme of lubricating the market economy. I say a "veil of frivolity" and "*seemingly* been subsumed" because it is critical to note that we continue to acquire and learn essential information from these "frivolous stories" (even if, or perhaps especially because, teaching is not their explicit purpose). For example, children are *always* learning from their parents even though there may not be any explicit or conscious teach-ing going on. Television may not make claims of educational programming except in specific instances, like *Sesame Street*, where education is the explicit goal, but as television tells its tales, learning is constantly taking place.

Storytelling has always been at the heart of how we make sense of the world and this is as true today, perhaps even more so, as at any point in history. As King queries: "Did you ever wonder how it is we imagine the world in the way we do, how it is we imagine ourselves, if not through our stories?" (2003: 95). That said, *the way* in which we tell and share stories has fundamentally changed very recently, as has the *content* of those stories. First, where throughout most of human history we shared stories face-to-face in intimate groups, most of us are now told most of our stories, especially our shared stories, by a few people via a few forms of mass mediated communica-tion. While there are other sources of shared stories, perhaps most notably religious stories, in terms of quantity of daily exposure, for most people, nothing touches the sheer volume of stories we receive from our mediated communication, especially television.

Thus we come to my third point about stories: mediated communication has allowed stories to be shared on a mass scale, and this is what gives mod-ern storytellers their great power. As academic and media activist Stephen Duncombe succinctly proposes: "Truth and power belong to those who tell a better story" (2007: 8). Of course we understand that stories alone do not confer truth, power and ruling status. The United States arguably tells more of the world's stories than any other country (because American media, and American television in particular, are very prominent in people's lives around the world), but the U.S. also relies on the world's best-funded army to rule. The concept remains, however, that the stories we are told play such a central role in how we understand reality that those who do the telling have tremendous power in our lives — and mediated communication gave them that power.

The Newness of Mediated Storytelling

One of the points I highlight with students is how incredibly new mediated communication is. For people born in the late twentieth century and early twenty-first century, mediated communication is — and always has been — so much a part of their lives that trying to convey how *new* this form of communication is in the context of human history is like trying to draw attention to the air they breathe. Or, as Marshall McLuhan offered: "It's like a fish in water. We don't know who discovered water but we know it wasn't a fish. A pervasive medium is always beyond perception" (McMahon and Sobelman 2002). That said, I believe it is important to at least try to convey just how new mediated communication is to humans and why this newness is important.

Sometimes I draw upon Carl Sagan's "Cosmic Calendar" in order to get at the idea of the newness of mediated communication. Sagan highlights what an incredibly recent addition humans are to the universe by using a twelve-month calendar as an analogy for the history of the universe. On the "Cosmic Calendar" humans arrive in the universe at 11:59 p.m. on December 31. So what happens if we use a similar analogy and create a calendar spanning the history of human communication?

According to the Smithsonian National Museum of Natural History, while our ancestors probably started talking to each other about 350,000 years ago, some of the earliest forms of mediated communication (any form of communication that does not require that the communicators be face-to-face, such as carvings) appeared about 40,000 years ago. The earliest forms of writing, using symbols to express words and concepts, appeared about 8,000 years ago. But the mediated communication technologies that defied time and space (the technology that today occupies so much of our lives) arrived much more recently. Alexander Graham Bell was granted the first telephone patent in 1876 — about the same time that personal use radios were first arriving. The late 1800s was also the time when motion pictures were first being created. Film with sound, or "talkies," appeared in the mid-1920s, at about the same time that television entered the public sphere. Some mark television's arrival with Philo T. Farnsworth's presentation in 1927 of a televised image of his wife. Depending on your age, that may seem not so long ago or unimaginably long ago, but it is only a few generations.

If we return to our analogy of a Human Communication Calendar, mediated communication appears at the end of the year and mass mediated communication appears in the final few minutes of December. As Robert Kubey and Mihaly Csikszentmihalyi point out:

> First human beings emerged on Earth approximately 2 million years ago. In this vast stretch of time, approximately 100,000 human generations have lived and died, and yet ours are among the first to live in a world where much of daily experience is shaped by widely shared instantaneous mass communication. (1990: xi)

In other words, we are creatures who have known, and relied upon, almost exclusively, face-to-face communication. Our brains, therefore, are used to

— and have evolved over thousands of years to — communicate face-to-face. Additionally, our brains have evolved over thousands of years to communicate with, and try to make sense of, not just each other but also the land and animals with whom we share the planet.

This idea of communicating with the land and animals isn't just romantic fanciful thinking; this is how we survived not so long ago. For most of human history, our survival was based on astute attention to our surroundings: if there was a movement in the bushes, for example, we had better react swiftly or, potentially, be eaten. That we have these "old brains" that have evolved to pay attention to such things has been studied extensively, most famously by Ivan Petrovich Pavlov. His research explored how we, and other creatures, have evolved to have conditioned reflexes (Pavlov's conditioning of dogs to salivate at the sound of a bell being his trademark example).

The implications of this evolutionary conditioning have not been lost on those who create television. Our instinctive, reflexive and involuntary attention to movement — and an automatic cognitive processing of that movement ("is that rustling in the bushes dangerous?") — is called an "orienting response," or an "orienting reflex," in the television research literature. While this orienting response is of great interest to the creators of television in general, it is of particular importance to advertisers. Advertisers are interested in the orienting response because they can use this to create ads that television viewers can't help but pay attention to. In a paper entitled "The Effects of Television Commercial Pacing on Viewers' Attention and Memory," researchers Paul Bollis, Darrell Muehling and Kak Yoon concluded:

> Given that skin conductance responses [how much a participant sweats] index viewers' arousal and their "orienting responses" to advertising, the current study provides some evidence that fast-paced advertisements (as compared to slow-paced advertisements) have the potential for enhancing involuntary attention to television advertising.... The subjects in this study recalled more advertisement-related bits of information when exposed to fast-paced advertisements than when exposed to slow-paced advertisements. (2003: 25)

This is all by way of highlighting the recency of television's arrival into our lives (and that this recency has implications); added to this, it is important to note the speed with which television has penetrated our lives. According to the University of Minnesota's mediated communication timeline, network television was established in the United States in 1949. Six years later, according to Mitchell Stephen's history of television, half of American homes had television. As I mentioned above, in 2008 Nielsen reported that 99 percent of American homes had at least one television set and more than half of American homes had three or more. Therefore television went from practically non-existent to close to a hundred percent in American homes in just over fifty years. I realize that compared with penetration rates of newer digital technologies, fifty years is an eternity, but from an historical perspective, television's story is an incredible one.

Television's relatively recent arrival is paralleled by the recent addition of television research to the more general field of "communication research." For example, the National Communication Association (NCA), one of the largest communication associations in the United States, had a focus on speech-related research well into the 1990s, only changing its name from Speech Communication Association in 1997. While there was some television research and publishing done in the 1960s (for example, Robert Lewis Shayon's *The Eighth Art*, considered one of the earliest examples of television criticism, was first published in 1962), it was not until the 1970s that the field of research known as "television studies" emerged.

Over these last few decades, however, television research has grown tremendously and continues to be the focus of much scholarly energy. One can find scholars who research various aspects of television at virtually every university in North America — and around the world — and their work is presented at international communication conferences. For example, in the first decade of the twenty-first century, the Communication and Mass Media Complete Database (a collection of about 770 communication and mass media academic publications) lists almost fifty thousand academic journal articles related to television. In what follows, I provide an overview of some of the ways our "old brains" have researched and tried to understand the implications for this new way of communicating.

Our Intuitive Knowledge of Television's Effects

I think that the most powerful argument for the influence of television is intuitive and based on our personal experiences. Ask any North American about their relationship with television, and there is *always* a relationship. Everyone has a relationship with television even if their relationship is that they did not watch when they were growing up (rare, indeed) or have never watched at all (rarer still); thus their relationship with television is all about how they are constantly explaining to people how it came to be that they have such an "unusual" relationship with television. I sometimes ask students to share what I call a "television autobiography." This is an exploration of the relationship the person has had with television throughout their life. I am always struck by the intimacy and passion with which the students tell their tales. Most have grown up with television and describe a relationship that is both social (involving viewing television with family members and, especially in the university context, viewing with friends) and solitary/intimate (often involving a television in one's bedroom, an increasing phenomenon I talked about above, and often includes the practice of regularly falling asleep with the television on). Pause for a moment: if I asked you to write a television autobiography, what would you write about? Can you remember when you first watched television? What was it like? Did you grow up with television? How did you watch? What were the feelings you associated with watching television when you were growing up? What about now?

Of course, not everyone grew up with television. What I am struck by in those few examples where television was largely absent in a student's childhood, however, is how aware they were of its absence and how they, very often, embraced its "late" arrival in their lives (feeling like they were making up for "lost time" with television). Similarly, older folks who didn't grow up with television are often very clear about when television did enter their lives and how their relationship with television evolved after that point. This is all by way of saying that no matter what your life's history with television, you have something to say, often very heartfelt, about your relationship with the medium.

It is worth noting that while my focus is North America, increasingly everyone *everywhere* can tell a story about their relationship with television. Even in parts of the world where electricity is scarce or non-existent, such as the isolated islands of Vanuatu in the South Pacific, where I started writing this book, people will cluster around a generator-run television or gather in a store that has a television. Even in the increasingly rare situation where someone has literally never laid eyes on a television screen, virtually every person in the world knows about television because they have come into contact with people who have watched.

Intuitively we know that we have a significant relationship with television and, at least implicitly, we know that we are affected by this relationship. Based on this fundamental understanding we have that television affects us, I highlight two versions of how we react when "confronted" with our relationship with television. I call one of the reactions "but television doesn't affect *me*" and the other "anecdotal acceptance of television's effects."

But Television Doesn't Affect *Me*

People can be very ready to talk about television in their lives, but some are skeptical when it comes to attributing any *real influence* to television. This reaction is along the lines of, "Sure we North Americans watch a lot of television, but that doesn't mean we're a bunch of dupes blindly accepting television's messages." Or, more likely, "that doesn't mean *I'm* a dupe." The interesting aspect of this response is how incongruous it is with our fundamental understanding of other socializing influences in our lives. Think of the matter-of-factness with which we understand the socializing influence of family, friends, school, work, sports teams, faith organizations and so on. We accept, without hesitation, that because of the sheer time spent with these people and in these places, we are affected, molded and altered (think of parents' common concern that their teenager's friends are a "bad influence"). We accept that much of who we are has to do with the "nurture" part of the nature-nurture equation (which is why parents, myself included, are often so concerned about everything related to our children: the "nurture" part of the equation matters). Yet people will often leave television out of the socializing mix or resist when television is put in the mix for them. Perhaps you are resisting right now. Would you add television into the list of primary socializing agents in your life? If not, why not?

Students who study communication are interesting in this regard. I have had many students say they are glad they study mediated communication because it can help them to sort of "guard against" the effects of television. While there is some truth to this, as media literacy education highlights (a topic to which I will return), it is also true that just because we know about the effects of mediated communication, this doesn't mean we can stop those effects. Think about this in the context of family and friends: we know they are also socializing forces, but an awareness of this doesn't stop their influence — even if for some reason (e.g., growing up in a dysfunctional family) we are *trying* to stop their influence.

From this "intuitive perspective," the case can be readily made for television being at least as influential as any of these other primary sources of socialization. Not only do we spend a tremendous amount of time exposing ourselves to television — increasingly, essentially from birth — we also often spend *more time* with television (and watch it with more regularity) than we spend time in these other socializing contexts. Keep in mind that with an average viewing of television at over five hours a day, only work/school and sleep take up more hours (and even that is not a given). But the case for television as a potent socializing force, on par with family and friends, includes another important argument: television is our *shared* storyteller. Family, friends, school, work, faith teachings and so on can provide quite varied, even conflicting, content to us, and those around us learn things from these contexts that are often very different from what we are learning. Throughout our lives the messages we receive from these sources can also change dramatically and be inconsistent.

Television, however, is remarkably consistent in the stories it offers us — across channels and throughout our lives. Think of the consistency of messages we receive about what it means to fit certain criteria in order to be attractive, or the desirability of consumption and materialism (a theme to which we return in the next chapter). Mass audiences, therefore, receive exactly the same content, and even if that content is not understood in exactly the same way, the essence of the content is processed similarly enough that huge numbers of us enjoy watching it. So while there may be only a handful of people who are our family members, friends and coworkers, there are millions with whom we share Thursday night prime-time television programming. All you have to do is start talking about characters and storylines from classic television shows like *The Simpsons*, *Friends* or *Seinfeld* to know that this is true — in North America, across generations, cultures and classes, we would all be able to join the conversation.

This raises another important point: television has also played a role in socializing these other socializing forces in our lives. Television is part of our family (our parents and siblings probably watch television), friends (our friends probably watch television), school (our school friends and teachers probably watch television), work (our colleagues and clients probably watch television), sports/clubs (our sports and club buddies probably watch television), church (our church leaders probably watch television), and so on. Television is not only a part of who we are, it is also a part of those around us. Television is in *all* of us; it is our great common denominator.

Therefore, to resist the idea of television's profound socializing role because "we're not a bunch of dupes blindly accepting television's messages" is like saying that "we're not a bunch of dupes blindly being socialized by our families and friends." Television's effects aren't about being duped or being smart enough to "resist" or being clever enough to decipher the messages. Television affects us in the same way that these other socializing influences do: through sheer time spent watching, exposure to certain stories and certain ways of making sense of the world — and we share these stories with those with whom we share our lives.

Anecdotal Acceptance of Television's Influence

Another common reaction to our intuitive knowledge of television's effects is a kind of broad, uncritical embracing of television's influence. While this is somewhat different from the above "but TV doesn't affect me" reaction, both reactions stem from our intimate relationship with television. In the first example, that relationship leads us to believe that we know the medium so well we couldn't possibly be profoundly affected by it. We come to our anecdotal acceptance of television's influence from a similar place of believing that our intimate relationship allows us to know the medium, and its potential for influence, extremely well.

A few years ago my father recommended that I read *Reviving Ophelia*. The book, by the well-regarded psychologist Mary Pipher, is a critical exploration of how American girls are socialized. One of the powerful socializing forces that Pipher draws upon as playing a negative role in these girls' lives is mass mediated communication. In one passage, Pipher offers the following:

> Cassie's been surrounded by media since birth. Her family owns

a VCR, a stereo system, two color televisions and six radios. Cassie wakes to a radio, plays the car stereo on the way to school, sees videos at school and returns home to a choice of stereo, radio, television or videocassettes. She can choose between forty channels twenty-four hours a day. She plays music while she studies and communicates via computer modem with hackers all over the country in her spare time. Cassie and her friends have been inundated with advertising since birth and are sophisticated about brand names and com- mercials. While most of her friends can't identify our state flower, the goldenrod, in a ditch along the highway, they can shout out the brand of a can of soda from a hundred yards away. They can sing commercial jingles endlessly. (1995: 243)

But it is television, in particular, about which Pipher is concerned. In another section Pipher talks about her own experiences watching the music video station MTV.

I was shocked by the sexual lyrics and scenes. In the first video, openmouthed and moaning women writhed around the male singer. In the second video, four women with vacant eyes gyrated in low- cut dresses and high black boots. Their breasts and bottoms were photographed more frequently than their faces. (1995: 34)

While Pipher is right to worry about the role television plays in the social- ization of these girls and young women, the real strength in making the connections lies not in our assumptions, or "anecdotal experiences," upon which Pipher almost exclusively relies, but in the television effects research. For example, Pipher could have drawn upon a study published in 1992 by Philip Myers and Frank Biocca. Myers and Biocca created an experiment in which young women aged eighteen to twenty-four were shown twenty-six minutes of television programming, including advertisements, created to showcase either "idealized" women's bodies, or "neutral" women's bodies. What the experiment demonstrated was that the women who were shown the idealized images were more likely to imagine that their own bodies were thinner. In other words, the researchers found that "body shape perception can be changed by watching less than 30 minutes of television" (1992: 126). Pipher could have then drawn upon extensive subsequent research, such as studies by Michael Levine and Linda Smolak (as highlighted in Thompson and Heinberg 1999), showing that this ability for women to "take on" televi- sion's idealized body images, but then coming face-to-face with the reality that they may *not be able to actually achieve the idealized body*, plays an impor- tant role in creating eating disorders such as bulimia and anorexia.

I discuss *Reviving Ophelia* not as a way of denigrating Pipher's book — on the contrary it is a moving and important work — but rather as a way of highlighting a common practice of making generalizations and assumptions about television's effects in spite of the fact that a wealth of television research exists upon which one could draw. Perhaps the best way to highlight such non-research-based conjectures about the effects of television is to frame what Pipher did differently. What if I, as a television researcher, wrote a book about psychology without any reference to the psychological research? I do not think the book would be published because my lack of knowledge about the field would be obvious. Claims about television's effects, however, seem to be treated differently. It is not only "non-TV" scholars who make broad generalizations about television. News broadcasters and social pundits often draw upon television to explain various social phenomena without any referencing of the literature that backs up those claims.

What follows is an overview of the field of television effects research. I offer this information with the understanding, highlighted above, that most of us already have a wealth of television knowledge gleaned from our lifetime relationship with the medium. I therefore present the formal theories and research as a way of providing a context for helping us make sense of those personal experiences and as a stepping stone to understanding television's broader societal impact. It can be easy for people to dismiss the significance of the socializing role that television has played ("but television doesn't affect *me*"), and conversely there can be a tendency to talk about television's effects with broad and vague generalizations ("anecdotal acceptance of television's influence"). Television studies is a rigorous academic field with important information that should be added to our own extensive and very personal experiences with television.

Television Studies

People have always been wary of mediated communication, and logically so. We are a species that has evolved relying almost entirely upon face-to-face communication. Mediated communication (any form of communication that does not require us to be face-to-face) has come at us with breakneck speed and profoundly affected society with each new addition. In the mid-1400s, Johannes Gutenberg's first printing project on his revolutionary movable type press was to mass produce the Bible — also known as the "Gutenberg Bible." The Bible has always been a mediated form of communication (given that it is a book), but until the printing press came along its production was a slow process of monks laboriously hand-crafting each page. Suddenly the printing press made the Bible more widely available, and its distribution no longer relied on the church (and its legions of monks). The new "mass reality" of the Bible decentralized religion, putting a once rare text in the hands of the masses — and it was all because of the printing press.

Marshall McLuhan's famous idiom that "the medium is the message" (1964, *Understanding the Media: Extensions of Man*) was a wise observation that each new communication technology, regardless of the actual "content" of that technology, brought with it related societal and structural changes. Elizabeth Eisenstein draws upon McLuhan when she points out in her article "The Rise of the Reading Public" that the printing press encouraged literacy, individualism, new group identities and scrutiny of leadership (1991). These broad social changes caused people a certain amount of fear and discomfort — emotions that have come to be associated with the arrival of every new form of mediated communication.

As with the printing press, each new communication medium has caused many social changes — and researchers have always eagerly embraced the opportunity to study those changes. Communication researcher Harold Lasswell (1902–1978) is credited with famously proposing that this wealth of mediated communication research all falls somewhere within the framework of "who says what to whom via what medium with what effect?" (1948). Communication research can be done on each element, or combination of elements, of *who* (the person/people sending the message) *says what* (the content of the message) *to whom* (the person/people receiving the message) *via what medium* (print? broadcast? digital?) *with what effect* (what happens as a result of the message?). Not surprisingly, there have been myriad approaches to answering the virtually infinite questions that rest within Lasswell's broad query.

There have also been myriad tweaks and critiques to Lasswell's question. For example, some researchers question the implicit linearity of Lasswell's maxim, noting that "noise" was missing (noise being the literal or figurative interference a message encounters as it travels from sender to receiver). All of this is by way of highlighting that mediated communication research, in its brief lifetime (essentially established in the latter half of the twentieth century), has explored a huge number of questions related to these relatively new forms of communicating. And no mediated communication research question has garnered more interest than "How should we understand our relationship with television?" While a wealth of research has been done on other mediated communication forms — newspapers, books, radio, film, the Internet, social networking — the medium of television has fascinated researchers most of all and continues to do so.

Why are researchers fascinated by television? I think the simple answer is because — as the statistics I highlighted earlier in the chapter attest — television, more than any other medium, is entrenched in our lives. Books, magazines, newspapers, radio, telephone and most recently digital communication technology have all been invited into our homes, all occupy and affect our lives, but television has consumed us and our lives the most.

Television occupied us when it first arrived. Robert Putnam notes in his 1995 article "Bowling Alone: America's Declining Social Capital" (a precursor to his hugely influential book *Bowling Alone: The Collapse and Revival of American Community*, published in 2001): "Time-budget studies in the 1960s showed that the growth in time spent watching television dwarfed all other changes in the way Americans passed their days and nights" (1995: 73). As the statistics I've previously offered indicate, television continues to occupy us.

So part of the explanation for why researchers have been fascinated by television is that television profoundly changed the flow of our lives through displacing time that we previously spent doing other things. Television was also intriguing because it brought strange new stories — and experiences — to us right in our homes. Suddenly there were people, places and situations in our living rooms that we had never experienced before or even thought about. So the challenge for researchers was to figure out how to study television such that they could begin to answer the huge looming question, "What is this thing doing to us?"

Television Researchers and Their Worldviews

Lasswell encourages us to understand that communication research can focus on the sender of a message, the message itself, the intended receiver of the message, the form or "channel" of the message, the message's effect — and endless variations and combinations of these elements. My focus here is on the tail end of Lasswell's question: with what effect? This area of television research is arguably the most common, although critical analyses of television's message, known as textual analysis, have also been a popular approach. The possibilities for other ways of exploring television are virtually infinite, and even within the study of television's effects, the possibilities for how to undertake the research are many.

As with all research, one way to categorize television effects research is by placing the approach along a continuum from quantitative (essentially research entirely based on numbers) to qualitative (essentially research without numbers). When I talk with students about this continuum, I point out that the way in which people think about and research television's effects tells us as much about the worldview of the researcher as the aspects of television that the person is trying to illuminate. There is no right or wrong way to ask questions or find answers, but there is a whole lot of personal preference. Let's start with the quantitative end of the television research continuum.

Quantitative researchers believe that the world, and thus television as well, is ultimately knowable through the application of the scientific method, or the numbers-based testing of phenomena. The essence of such testing involves creating an "experimental group," which is given some kind of exposure or treatment (e.g., participants watch an advertisement) and a "control group," which is not given the exposure or treatment (e.g., participants don't watch the advertisement). If the experimental and control groups are essentially equivalent except for the exposure or treatment (usually achieved through a random assignment of participants to each of the groups), then differences that occur after the exposure or treatment (e.g., a greater desire of the experimental group to buy something) can be attributed to the exposure or treatment.

There are some fundamental rules for using the scientific method. First, one must make predictions. One usually draws upon previous research (often in the form of a theory) to predict or hypothesize what will happen once the exposure or treatment occurs. Second, it must be possible to fail; there must be a way for one's predictions or hypotheses to be refuted. That is, the theory cannot be so broad and sweeping that it is always right simply because it is impossible for it to be wrong. Third, it must be possible to repeat the research. The steps taken to test the predictions or hypotheses must be clear enough that someone else could follow the same steps, and when someone does follow the same steps, the same results should be achieved.

We tend to associate the scientific method with the hard sciences — chemistry, biology, physics. But other fields, including communication, draw upon the tenets of the scientific method to test social phenomena. When television effects researchers make use of the scientific method to test hypotheses, they are "social scientists." For television scholars the quantitative approach to research usually involves the testing of a theory (i.e., testing whether what a communication theory proposes should happen in a particular situation is, in fact, what happens).

Television effects researchers sitting at the far quantitative end of the continuum therefore embark on research that is similar to work done by their hard science counterparts in that their research is predictable, testable, repeatable and refutable. Inherent in this approach to research is also a belief that phenomena are observable and measurable. This type of research is called quantitative research because the findings of the research can be calculated based on numbers (e.g., numeric scores from survey questions, response times to certain activities, physiological indicators such as changes in heart rate or quantity of sweat, etc.). In addition to the importance of numbers, quantitative researchers also tend to value objectivity (i.e., that a particular researcher had no impact on the outcome of the research — thus harkening back to the importance of repeatability: findings aren't supposed to be dependent on a specific researcher or a unique set of circumstances).

It is important to note that while quantitative researchers tend to think of their work as being, at least ideally, objective and neutral (i.e., the phenomena being tested "speak for themselves") adherence to the scientific method comes with its own, often implicit, values and beliefs (stories, if you will) about the world. In other words, by reducing phenomena into isolated components and quantifying outcomes, there are inherent values and beliefs about the way the world works (e.g., that the world broken into isolated components can speak to the infinitely more complex whole).

At the other end of the television effects research continuum, we find researchers who believe that we cannot force complex human experiences into neat clean numbers. Relationships between television and its viewers are too messy for quantifiable outcomes, these researchers believe, and thus the research needs to be done using more subtle, and less scientific or numeric, approaches. Qualitative researchers feel, in essence, that the world is a chaotic and complicated place — and people are individuals whose experiences of the world cannot be averaged in with others' experiences. As such, the demand by quantitative researchers to create testability, predictability and refutability leads to artificially simple situations and relationships. Therefore, instead of testing the impact of an advertisement with experimental and control groups (as proposed above) qualitative researchers might ask participants to watch an advertisement and "think aloud" everything that comes to mind. The researchers could then explore common words and themes in participants' descriptions of what they were experiencing.

Another way of distinguishing quantitative and qualitative research is by looking at where each type of research tends to occur. The location of choice for many quantitative social scientific studies is a laboratory. A lab is an ideal environment for scientific research because the objective of isolating a particular relationship for study is easier to achieve there (i.e., in a lab various elements can be controlled while the variable that is being tested can be manipulated). In "the field" (pretty much any location other than a lab), various elements — sounds, weather, other people, etc. — may affect the research in ways that are not in the control of the researcher. But "the field," qualitative researchers would say, is more reflective of the real, complicated, messy world that we inhabit and, as such, is where data should be collected.

Qualitative researchers, like their quantitative counterparts, do try to be rigorous in their theorizing of the world and their establishment of criteria for studying phenomena. For example, while there are not the same requirements for prediction, refutation and repeatability as found in quantitative research, qualitative researchers often make a case for their research based on how their findings make sense to those who participate in the study. A qualitative researcher may also actively try to locate examples that contradict their findings as a way of evaluating the robustness of their research. One qualitative approach to studying television effects, therefore, has been to observe viewers in the "real world" and to reflect on what appears to be going on as "real people" watch television in "real situations." Once this data has been collected, the researchers can return to the participant viewers and ask if the conclusions resonate for them.

A subset of the qualitative researcher category is the critical-qualitative researcher category. These researchers share qualitative researchers' notion that the relationships between television and its viewers are messy and complicated. In addition, critical researchers see television as being involved in the creation and maintenance of inequitable power distribution within society. Critical television researchers, therefore, believe that their research not only illuminates the ways in which television creates and maintains societal power imbalances, but their research often also addresses how to aid those who are on the losing side of the imbalance. A critical-qualitative study of advertising might involve a focus group in which economically disenfranchised youth are encouraged to talk about how advertising affects them and are then given the opportunity to create alternative messages for social change. The researchers could describe the process by which deconstructing mainstream advertisements became the foundations on which the alternative ads were constructed.

Critical researchers, like qualitative researchers, therefore steer clear of concepts like prediction, refutation and repetition (in fact, critical researchers would point out the way in which such approaches to understanding the world maintain the status quo). Critical researchers do, however, have criteria for quality work, including how the critical analysis and suggestions for reform speak to others who are engaged in similar analyses and who share an interest in reform. Critical thinkers have generally taken the position that television is problematic — because of the content, time displacement, etc. — and have often made a case that society should be taking television's role much more seriously or, in some cases, should do away with television altogether.

While the quantitative, qualitative and critical-qualitative research approaches are, at their core, quite different from each other, they are by no means mutually exclusive. Placing the approaches on a continuum highlights that the research approaches overlap and blend into one another. In fact, in a "mixed methods" research design, a researcher consciously applies different methodological approaches to the same question with the hopes that the weaknesses of one approach will be shored up by the strengths of the other approach. Joanna Sale, Lynne Lohfeld and Kevin Brazil point out the following in their study of mixed methods:

> Qualitative and quantitative research methods have grown out of, and still represent, different paradigms. However, the fact that the approaches are incommensurate does not mean that multiple methods cannot be combined in a single study if it is done for complementary purposes. (2002: 50)

What's the Best Way of Doing Television Research?

While communication researchers are everywhere along the quantitative-qualitative continuum and may undertake research that makes use of both approaches in the same study, it is also true that they tend to fall into fairly rigid methodologically defined camps. Certain communication researchers believe that unless research is about quantifiably testing a theory, it is not research that is worth doing. Indeed, some communication journals require a paper to contain theory-derived hypotheses that have been tested in order for the paper to even be reviewed. (I have had a paper returned to me, unread, with exactly this reasoning.)

Others believe that quantifiable research is the "easy way out" because number-crunching avoids the complicated subtleties of the real world. I was once at a conference where Neil Postman, the author of *Amusing Ourselves to Death*, joked that at New York University, where he was a professor, the "weak students" were encouraged to do quantitative research. Thus, there is tension between researchers who espouse these differing worldviews. While many would argue that all research is valuable, there is an ongoing (if often implicit) struggle over what is the "best" way to approach television effects research.

Given this struggle, it is particularly telling that television effects researchers have come to at least two shared fundamental conclusions: television's stories are important; television's stories influence us. In what follows, I lead the way through some of the television effects history, theories and research in order to provide a context for evaluating the impact that television's stories are having on us.

The History of Television Effects Research

Magic Bullet/Hypodermic Needle

One of the important things to keep in mind about television, and mass mediated communication more generally, is how incredibly new it is. Yet the research into these "brand new" mass mediated forms of communication has seen many changes over a relatively brief period of time. The earliest studies of mass mediated communication have been given the name "extreme effects," or "massive effects," research and the early theory had similarly dramatic names, like "magic bullet" or "hypodermic needle." This theory was essentially based on the fact that for millennia we lived in a world where what we saw was what we got. Our quick reflexes based on what we saw, smelled or heard allowed for survival: Rustling noise? Run! Glimpse of deer? Draw bow!

Mediated communication changed everything — extremely and massively. Suddenly a predatory animal could appear (on a screen) and not be real (at least not in the flesh and blood sense). What the magic bullet, or hypodermic needle, theory proposed was that you and everyone else with these millennia-evolved "old brains" would respond as though that predatory animal on the screen was *real*. The example I use in teaching media effects, which is probably apocryphal, is of an early movie theatre showing a film that contains the sound and image of a train coming into the screen. The audience, en masse, gets up and runs out of the theatre. Another version of this story involves a movie of a cowboy shooting directly at the audience and audience members diving to the floor to avoid the bullets.

The earliest approach to media effects research, therefore, proposed that what we received via the media was like a magic bullet or a hypodermic needle because we would all, being old brain creatures, react to, and be affected by, mediated communication in the same instinctual way. The massive effects research involved both quantitative and qualitative studies and explored how exposure to mediated messages played out, directly, in people's lives.

Many who teach about media effects point to a radio broadcast of H.G. Wells' *War of the Worlds* on October 30, 1938, as representing the end of the magic bullet theory. The *War of the Worlds* tells the story of Martians landing their spaceship in New Jersey and proceeding to invade Earth. (The same story was turned into a movie starring Tom Cruise in 2005.) As the radio broadcast aired, two things became clear. First, some people panicked because they believed the story events were really happening. Second, and even more interestingly, at the time not everyone panicked because not everyone believed the broadcast was real.

To us, this seems like a logical, and obvious, combination of audience responses: some people are more gullible than others so some reacted and others did not. But in an age of massive effects thinking, the audience's heterogeneous response was fascinating. A researcher by the name of Hadley Cantril seized the moment and questioned what the differences were between those people who believed the broadcast was true and behaved accordingly (e.g., panicked) and those who took measures to establish the fiction of the program (e.g., called the radio station).

Cantril found, for example, that those with more formal education were more likely to undertake activities like calling the radio station in order to verify the situation. Cantril's study, published in 1940 in a book entitled *The Invasion from Mars: A Study in the Psychology of Panic*, and other subsequent studies like it, established that there was no magic bullet or hypodermic needle. Rather, mediated communication audiences, like all groups of people, are composed of diverse individuals who interact with what they are receiving in diverse ways.

The idea, and often fear, that mediated communication could have profound effects on large numbers of people, however, continued as a driving impetus for research. In no area of research did this idea manifest itself more vigorously than in the possible relationship between children's exposure to television and anti-social (i.e., aggressive, violent) behaviour. Concerns about such links, and anecdotal reports of children modelling aggression and violence that they had viewed on television, led to the creation of the U.S. Surgeon General's Scientific Advisory Committee on Television and Social Behavior in 1962.

In 1972, the Advisory Committee released their results in a five-volume report. The report contained findings that were both cautionary (people consume a lot of violent television and there do seem to be links between viewing violence and subsequent aggressive/violent behaviour) and qualified (the aggressive/violent behavior is a result of very specific conditions). A summary of the Advisory Committee's findings offers some tentative conclusions and something close to an apology that the results were not more definitive.

> [The researchers found that there is] a preliminary and tentative indication of a causal relationship between viewing violence on television and aggressive behavior; [there is] an indication that any such causal relationship operates only in some children... and an indication that it operates only in some environmental contexts... Such tentative and limited conclusions are not very satisfying. (1972: 11)

The death of magic bullet-type thinking is held up as what is significant about the *War of the Worlds* broadcast and Cantril's research. Subsequent studies, such as the Surgeon General's report, continued to highlight that the media's effects seemed to be less massive and more complex than previously thought. But what was at least as important, and could get somewhat lost as researchers moved away from massive effects thinking, is that *mediated communication was affecting people* (just not in the linear indiscriminate ways once imagined).

Yes, the *War of the Worlds* radio broadcast was significant because not everyone panicked, but of course what is also true is that because of one fictional radio story about Martians, a whole bunch of people *did* actually believe that Martians had invaded Earth and responded as such. Many other people were affected enough to at least look into the situation. From an effects perspective, that's still pretty powerful stuff! So, while the magic bullet theory proved to be a simplistic way of thinking about mediated communication effects, aspects of that theory (mediated stories affect us and our understanding of reality) continue to be vital in our ongoing researching and understanding of media effects.

Minimum Effects

After the death of magic bullet theory, the theoretical pendulum swung away from massive effects and towards what is sometimes referred to as "limited effects," or "minimum effects." This understanding of our relationship with mediated communication focused on just that: the relationship readers/listeners/viewers have with mediated communication. Limited effects is in many ways the antithesis of magic bullet thinking. Instead of being bombarded and defenceless, we are in charge. Indeed, the language of effects became downright wishy-washy — along the lines of "under certain circumstances some media may have some influence on some people some of the time." Of course, for those of us peering back from our perch in the twenty-first century, this more cautious and nuanced understanding of how we might be affected by mediated communication makes sense, but at the time the realization was profound.

The best known and most researched minimum effects communication theory is "uses and gratifications." The essence of this theory is that media use, and media effects, are all about the individual. Each individual actively seeks out media content and makes choices motivated by such things as interests, needs and values. Mediated communication provides messages and experiences, therefore, that are specific to the individual. So, as you can see, this really is fundamentally different from magic bullet thinking. Mediated communication isn't all powerful, but instead people make choices about, and are affected by, their media consumption in the same kinds of ways that they make choices about and are affected by something like, say, their food consumption. For example, uses and gratifications theory proposes that the way we are affected by the media depends on who is doing the consuming. Taking the example of food consumption, some people ravenously seek out and derive great pleasure from eating meat, even raw meat. Others never eat meat and may find even the sight of it repugnant.

Research guided by uses and gratifications theory asks individuals what types of mediated communication they seek out to fulfill what needs. For example, in the early 1940s, Herta Herzog interviewed individuals who listened to serial radio programming and was able to place their uses and gratifications into three categories or factors: emotional release, wishful thinking and seeking advice. In 1949, Bernard Berelson asked people during a newspaper strike what they missed about reading the newspaper and found five uses and gratifications: information, interpretation of current affairs, tips for daily living, social prestige/contact and relaxation. Early television uses and gratifications research undertaken by Denis McQuail, Jay Blumler and Joseph Brown found that "people are motivated to watch television for: diversion — to escape and for emotional release; personal relationships — for companionship and social utility; personal identity — for personal reference, reality exploration and value reinforcement; and surveillance — to acquire news and information" (1972: 530). In 1974, Bradley Greenberg explored children's and adolescents' television uses and gratifications and found seven main reasons for viewing: habit, relaxation, companionship, pass time, learn, excitement and escape.

Again, while you might scratch your head and think, "Well, duh, I could have told you that people seek out television viewing for different reasons," it is important to keep in mind the historical context out of which this research was being done. Uses and gratifications theory not only acknowledged that there are effects — such as relaxation, escape, learning — from television, but it also highlighted that those effects are personalized and that it is the viewer who chooses the content and, essentially, seeks out the effect.

What uses and gratifications theory and magic bullet theory have provided for us can be understood as bookends, such that the majority of theories about television's effects fall somewhere between the two. However, it is important to note that while uses and gratifications theory is alive and well in this age of exploring how/why people use digital communication, nobody would actually do research in the name of magic bullet theory any more.

In what follows, I provide a brief overview of a few of the television effects theories that sit somewhere between these two bookend theories. Of course there are many more television effects theories than I have space for here (in fact each of these theories could easily occupy volumes and in some cases they have), but what I share are the theories that I think are most interesting and helpful in creating a context for my case against television.

Social Learning Theory

In a classic study known as the "bobo doll experiment," psychologist Albert Bandura explored the circumstances under which children would imitate an adult's violent behaviour towards a bobo doll (a blow-up, clown-like doll that can be knocked over but rights itself). Bandura and his colleagues Dorothea Ross and Sheila Ross found that when a child watched an adult beat up the bobo doll (sit on the doll and punch its nose, hit the doll on the head with a mallet, throw the doll in the air and then kick it) and was then left alone with the same bobo doll, the child "readily imitated the aggressive behavior" (1963: 3).

Bandura and his colleagues extended the modelling of violent behaviour to include a television version (the same adult abusing the bobo doll in the same ways) and a cartoon version of "Herman the cat" abusing a bobo doll. Would the fact that some of the children were learning behaviour via a screen matter? Yes. The authors explain: "The available data suggest that, of the three experimental conditions [human in the same room, human on television, cartoon cat on television], exposure to humans on film portraying aggression was *the most influential in eliciting and shaping aggressive behavior*" (1963: 7, italics added). Not only did the children learn aggressive behaviour from the screen, they learned aggressive behaviour *better* from the screen than they did from someone in the room with them!

Based on these early studies and extensive subsequent research, Bandura developed social learning theory. This complex theory went on to incorporate variables such as the role of reward and punishment in how we learn, but for our purposes the essence of the theory is most important: we learn from television. I know you're not surprised by this conclusion, but it is a fundamental and foundational finding that is important to keep in mind as I share the following theories with you.

Dependency Model

Have you ever hung out with a famous movie star? Sat down with the president of the United States or the prime minister of Canada? Most of us haven't, but that doesn't stop us from knowing a whole lot about movie stars and politicians. This is the essence of the dependency model. Sandra Ball-Rokeach and Melvin DeFleur proposed in their 1976 *Communication Research* article "A Dependency Model of Mass-Media Effects" that the media have the most impact when they tell us about things, and people, we haven't experienced for ourselves. When I first encountered the theory, I was immediately drawn to it because, at its core, the dependency model echoes some of what I have proposed about the importance of storytelling in our lives.

As I discussed earlier, throughout human history stories have played an essential role in how we have understood ourselves and our world, and until recently, these stories were told to us by those with whom we lived, and the stories provided us with critical information. In Joseph Campbell's terms, these were stories that told us about the metaphysical and cosmological — such as creation stories and stories about what happens after death — as well as the sociological and pedagogical — such as the day-to-day practicalities of how to predict the weather, look after children, hunt and gather.

Our storytellers had important experience relevant to our lives; we trusted them to tell us about what we needed to know. Ball-Rokeach and DeFleur's dependency model highlights a fundamental shift in storytelling that occurred during the Industrial Revolution (mid-eighteenth to mid-nineteenth centuries), when the forces of migration, industrialization and urbanization changed our lives in two fundamental ways. First, these forces uprooted us — removing many from families, communities and trusted storytellers — and placed us in newly forming urban centres. Second, these forces created the fertile context out of which early mediated communication such as newspapers and magazines arrived (i.e., mechanized presses created the publications, dense population allowed for easy distribution and increasingly literate individuals longed for stories about their new lives).

Therefore, suddenly (in the context of human history) the essence of how we acquired information about ourselves and the world in which we lived shifted. In the blink of an eye, large numbers of people were not in small, tightly knit communities. People were not living in the same place that generations of their ancestors had lived. People therefore needed new information, new stories, about their new lives. "Persons living in societies undergoing change from traditional to industrial forms experience pervasive ambiguity. This ambiguity is particularly acute during the period between their psychological unhitching from traditional customs, values, and world views and their adoption of more modern versions" (Ball-Rokeach and DeFleur 1976: 10). And who told these "more modern versions" of stories about "customs, values and world views"? Mediated communication. Newly arrived in urban, modern and industrialized societies, people became dependent on mediated communication to tell stories and contextualize their lives.

The dependency model proposes that mediated communication affects us most when we have no personal context for what we are being told and are, therefore, entirely dependent upon the storyteller. While it is perhaps easiest to see how dependent we were on mediated communication in the historical context of the Industrial Revolution (i.e., at a time when we were suddenly removed from our roots, our community and the land), our reliance on mediated communication's stories has been ongoing. We have depended upon, and continue to depend upon, mediated communication to tell us about who we are and the world in which we live. Turn on the news. How many of the stories you're being told do you know about from personal experience?

One way that researchers explore the dependency model is by measuring whether higher levels of media dependency are related to media effects. This requires researchers to quantify how dependent people are on the information they receive from a medium like television and to quantify the effects of that dependency. Sandra Ball-Rokeach, Milton Rokeach and Joel Grube (1984) developed a Likert scale to measure media dependency for television. The scale has also been used to measure dependency on media other than television. For example, in 2002, Padmini Patwardham and Jin Yang used the scale to explore Internet dependency. This scale asks participants a series of questions, each of which starts with, "In your daily life how important is television to…" and participants can respond from 1 (not at all important) to 5 (very important). Here are the eighteen questions:

1. Stay on top of what is happening in the community?
2. Unwind after a hard day or week?
3. Gain insight into why you do some of the things you do?
4. Discover better ways to communicate with others?
5. Decide where to go for services such as health, financial, or household?
6. Relax when you are by yourself?
7. Find out how the country is doing?
8. Imagine what you'll be like as you grow older?

9. Give you something to do with your friends?
10. Figure out what to buy?
11. Think about how to act with friends, relatives, or people you work with?
12. Have fun with family and friends?
13. Observe how others cope with problems or situations like yours?
14. Keep up with world events?
15. Be a part of events that you enjoy without having to be there?
16. Get ideas about how to approach others in important or difficult situations?
17. Plan where to go for evening and weekend activities?
18. Have something to do when nobody else is around?

The second part of dependency model research involves exploring whether a participant's higher level of media dependency (i.e., a higher score on the above scale) is related to stronger media effects (e.g., longer and more accurate retention of mediated information, greater willingness to act on the mediated information that one receives, etc.). The answer, in numerous dependency studies, has been "yes." For example, in 1979, Ball-Rokeach, Rokeach and Grube were interested in testing the dependency model in a situation that would, in as many ways as possible, mimic what takes place in the real world. The researchers designed and produced a half hour television documentary called *The Great American Values Test* as the foundation for their study. *The Great American Values Test* documentary promoted the themes of egalitarianism (especially as it relates to women and people of colour) and environmentalism. The three major American television networks of the day (ABC, CBS and NBC) agreed to run the documentary at the same time (7:30–8:00 p.m.) on a weekday evening, and its airing was promoted in print and television ads. In other words, everything about the documentary was "real," including that a Hollywood actor (Ed Asner) and a news anchor (Sandy Hill) hosted it.

As soon as the documentary was over, a sample of residents was contacted and asked whether they had watched all or part of the show, if anyone else in the home had watched it, their reaction to it and some demographic questions (gender, age, etc.). A couple of weeks after the program aired a sample of residents in the broadcast area received a survey that measured their media dependency as well as various attitudes, values and beliefs. A sample of residents who had not received the television broadcast was also sent the survey. The final aspect of the study was a mailed solicitation for funds from organizations related to the egalitarianism and environmentalism espoused in the documentary. Based on responses to this solicitation for funds (i.e., whether people made donations and in what quantity), the researchers concluded that there had been a dependency effect. By comparing those who viewed the documentary in its entirety and who were high in dependency for the documentary's messages, with those who were low in dependency or had not seen all or any of the documentary, the researchers were able to conclude that high dependency individuals who had viewed the entire documentary were more likely to make a donation in response to the mailed solicitation. The dependency model was therefore accurate: the documentary had more of an effect on those who had greater dependency on its message.

What I find most appealing about the dependency model, however, isn't how it has "performed" when tested (although the reason for its survival is that it has performed as predicted when tested), but rather the support for intuitive logic it provides in a discussion of media effects. The dependency model proposes, at least implicitly, that in the time before mediated communication, and throughout most of human history, our understanding of the world was narrower than it is today. We tended to only know about that which we had personally experienced or had been told about by someone close to us with personal experience. But our understanding of the world was also deeper, and richer, in many ways than it is today.

This deep, rich knowledge is perhaps nowhere more apparent than in the relationship we had with the land on which we lived. Our lives were dependent on knowing the nuances of weather patterns, seasons and growing cycles. Our lives were also dependent on being aware of and knowledgeable about the other life forms with whom we shared the land. The stories, spirituality and teachings of the world's indigenous peoples often exemplify this knowledge, dependence and respect related to the planet and its inhabitants. (Jerry Mander talks about this quite extensively in his book *In the Absence of the Sacred: The Failure of Technology & the Survival of the Indian Nations.*)

When we began the exodus from the land and from our small family-based communities (an exodus that continues to this day), we traded in our narrow, but deep, knowledge for a kind of broad, shallow knowledge. Note that I am not saying we know more or less than we used to, but I am saying that we know *differently*. The things we used to know — about the acquisition of food, the building of shelter, the patterns of the weather, the movement of the seasons, the habits of animals, etc. — we knew from our direct experiences with seeing and doing and from receiving stories from those in our lives who had direct experience seeing and doing.

Today we have unbelievable and unprecedented access to information. We certainly have more formal education than our ancestors, and our range of knowledge is exponentially greater. But the vast majority of what we know has not come from direct experience. Instead we have become dependent on television (and other mediated communication) to tell us about far flung parts of the world (most of which we have never visited), other people (most of whom we have never met), images (many of which we have only ever seen on a screen) and about values, attitudes, beliefs and "desirable behaviours" we might never have come up with on our own.

The people from whom we receive television's tremendous information are not our family members who care for our emotional, spiritual and physical well-being, but instead they are television executives and personalities who have an almost exclusive interest in selling our eyeballs to advertisers. It is these executives, and those who work for them, who choose *which* stories you will be told (from amongst the virtually infinite possibilities) and *how* those stories will be told. This is exactly what the next theory focuses on. In other words, while the dependency model highlights the general importance of our *reliance* on television's stories, what I appreciate about the next theory is that it encourages us to focus on how television's stories are selected and told.

Agenda-Setting

When students are introduced to agenda-setting theory, the name of the theory causes them to think "conspiracy." While there are fundamental questions about power and influence contained within this theory, it is not about conspiracy (as such). There are no dimly lit back rooms where media moguls quietly plot to impose their agenda on the people. Agenda-setting theory begins with a more benign notion that at any moment on the planet there are, essentially, an infinite number of things going on. At a personal level we encounter this barrage as we go about our daily lives and constantly filter this input so that we are not overwhelmed by the endless sights, sounds, smells, tastes and emotions.

Richard McCombs and Donald Shaw drew upon this idea of the filtering of reality as they developed agenda-setting theory. They realized that in the same way that we as individuals must select that to which we pay attention, so too do those who create mediated stories. There are only so many stories, of whatever form, that the media can pass along to us; somehow decisions must be made about which stories to tell and how to tell those stories. McCombs and Shaw refer to the people who make these decisions about television's (and other media's) stories as "gatekeepers."

The setting of television's agenda by these "gatekeepers" (a category that includes television owners, producers, editors, storywriters) is therefore an essentially unselfconscious process. The process has much less to do with conspiracy and much more to do with subjective decisions about what are the "important stories" to tell when the underlying goal is selling the largest possible audiences to advertisers. Once gatekeepers have determined what the important stories are, agenda-setting theory proposes that these stories become *our* most important stories: television's agenda becomes the television viewer's agenda.

You may be wondering how television conveys the importance, value or "salience" (in the language of the theory) of what is being shared with us. In other words, when you watch the news on television (and much of the agenda-setting research has focused on television news), how do you know what the most important stories are? When I ask students this question, they only need a moment before they start to offer up the "ingredients of importance." My guess is that you have some ideas as well. You may have thought of such elements as frequency (i.e., how often a news item is in the news — the more frequently the story appears, the more important it is); location (i.e., where the story airs in a news lineup — the "top story" is usually just that, the first story in a broadcast and the one that is considered most important); time (i.e., how much time is spent on the story — the more time spent on the story, the more important the story); messenger (i.e., who tells the story — a reporter sent specifically to cover a story on "special assignment" is more important than the news anchor simply telling us about something); production value (i.e., the higher the production value in terms of images specific to the story, computer graphics, etc., the greater the importance); and many other elements.

One aspect of agenda-setting research has, therefore, been the development of criteria with which to ascertain a ranking of the "media's agenda" (i.e., how to go about answering the question "What are the top five most important news stories?"). Different agenda-setting studies have created different criteria, but the studies have tended to draw upon the kinds of elements highlighted above. The other aspect of agenda-setting research has involved developing the criteria with which one can ascertain what consumers of the news feel are the top news stories. This measuring of the public's ranking of importance is usually done by asking (e.g., in a survey) research participants to rank the order of the issues they think are important or by relying on existing data (e.g., public opinion polls undertaken by the government).

McCombs and Shaw undertook the first agenda-setting theory research during the 1968 U.S. presidential campaign. By analyzing how nine media outlets (five newspapers, two magazines and two television networks) prioritized the campaign's issues and by comparing this to the public's prioritizing of issues, the researchers were able to show a strong parallel or correlation between the media's and public's priorities or agendas. McCombs and Shaw argue: "In choosing and displaying news, editors, newsroom staff and broadcasters play an important part in shaping political reality. [Media consumers] learn not only about a given issue, but also how much importance to attach to the amount of information in a news story and its position" (1972: 176).

Since this initial study, hundreds of agenda-setting studies have been undertaken, many of which have focused on elections. There have, however, been many other contexts in which agenda-setting research has taken place (such as for the priorities within a specific issue like the environment) and for many different types of media (such as new digital forms of communication). In 2006, Wayne Wanta and Salma Ghanem published a statistical overview, or meta-analysis, of agenda-setting research across contexts and mediated sources of information. Their findings confirm that we are, indeed, affected by the agenda-setting function of the news media; what the news tells us is important, we come to believe is important.

You may be wondering about the influence of first-hand knowledge — what happens when there is a news story that we have personally experienced? Presumably we would not rely as heavily on television's information in that case. When we already have a lot of knowledge about a story, we have a filter through which we can judge for ourselves whether the story is important or not. Agenda-setting theory uses the term "obtrusive" for those issues with which we have personal experience (for example, a community's garbage disposal crisis is likely an obtrusive issue for the community members who are increasingly surrounded by garbage) and "unobtrusive" for those issues with which we do not have personal experience (for example, a garbage crisis in a far away land).

Television's, and all media's, agenda-setting effect has been shown to be stronger for unobtrusive issues because (as dependency theory also highlights) we are completely reliant on television to tell us about the unobtrusive issues. Similarly, while we might not pay much attention to television's news stories that have no relevance for us, we will pay particular attention when we encounter stories that we are personally involved in. Therefore, much like dependency theory, agenda-setting theory proposes that television affects us the most when we are completely reliant upon it to tell us something relevant, for example, news about a far away war in which a family member is fighting.

How do we know that it is television's agenda that affects us and not our agenda (the issues that the public feels are important) that affects television? McCombs and Shaw addressed this question in their original 1972 article. They pointed out that in order for the public to influence the mass media's agenda, or for the public and the mass media to somehow "independently" arrive at the same ranking of presidential election priorities, the public would have to have access to information *other than* what they received via the mass media. "Since few [people] directly participate in presidential election campaigns and fewer still see presidential candidates in person," McCombs and Shaw offered, "the information flowing in interpersonal communication channels is primarily relayed from, and based upon, mass media news coverage" (1972: 185).

Others have gone about specifically researching the possibility that it is the public's agenda that affects the media's agenda. For example, in their overview of agenda-setting research, Maxwell McCombs and Amy Reynolds highlight G. Ray Funkhouser's study of public opinion and media content throughout the 1960s, which found:

> There was no correlation at all between the trends in news coverage of major issues and the *reality of these issues*. But there was a substantial correlation… *between the patterns of news coverage and the public's perception of what were the most important issues.*" (2002: 6, italics added)

Funkhouser, therefore, concluded that the news media create a sense of what is real and important in the world and not the real world itself.

A final important aspect of agenda-setting theory begins with Bernard Cohen's now immortalized offering in his 1963 book *The Press and Foreign Policy* that the media "do not tell us *what* to think, but they do tell us *what to think about.*" The media can provide us with the "important" stories but they cannot tell us how to make sense of those stories. Only we can decide how to make sense of the stories we're told. Right? There is, however, another part of agenda-setting theory that challenges Cohen's assertion that we think what we want about the stories we're told.

Much of agenda-setting theory research involves researchers investigating how the media's valuing of an issue, or in the language of the theory, the "salience of an object," is conveyed to the public. This salience is conveyed based on television's telling of a story (placement, length, formatting, images, etc.). More recently, researchers have added a second level to their analysis by investigating *how* a particular mediated story is told (i.e., not just how a story's importance is conveyed). This research process, often referred to as framing analysis, asks questions such as: What aspects of a story are selected? What is highlighted within the story? What is elaborated upon? What is excluded?

In his book *The Whole World Is Watching*, Todd Gitlin explores how the media framed the coverage of police crackdowns on the 1960s student uprisings. He offers: "Media frames are persistent patterns of cognition, interpretation, and presentation, of selection, emphasis, and exclusion, by which symbol-handlers routinely organize discourse whether verbal or visual" (2003: 7). It is not only, therefore, that stories "appear" to us in certain ways (placement, length, formatting, images) but also that those stories are *told to us* in certain ways. Therefore, in addition to Cohen's early and prescient observation that the media tell us what issues to *think about*, the framing of those issues means that we are also being told *what* to think.

Dependency theory and agenda-setting theory leave us with a substantial body of research that points clearly to the conclusion that while television may not be a magic bullet, it is a powerful influence in our lives — especially when television tell us things about people we have never met, places we have never been, situations we have never encountered and so on. Or, in the language of the theories, we are particularly affected by mediated messages when we are dependent on them and when they involve unobtrusive, yet relevant, issues for us. However, because there is no magic bullet, not everyone comes to believe in, or agree with, television's version of reality. The next theory explores the implications for disagreeing with television's reality.

Spiral of Silence

What happens when a "minority of one" is pitted against a "unanimous majority"? This was the question sociologist Solomon Asch addressed in the 1950s with a series of experiments that involved assessing the length of lines. While there were many experimental variations, the essence of Asch's research involved telling a research participant that they and a small group of others in the room were part of a "vision test." The research participant was then shown two cards. On one card was a single line and on the other card were three lines. The participant was told that the goal was to match the single line on the one card with the line of equal length on the second card.

All of the other people in the room were "undercover" confederates involved with the study. These undercover confederates would offer their answer before the actual research participant and unanimously chose the wrong line. Asch found that "seventy-five per cent of experimental subjects agree[d] with the [incorrect] majority in varying degrees" (1955: 4). In other words, in spite of the fact that the research participant knew that the line being chosen as the matching line was incorrect (it was quite obvious), he or she went along with the others and chose the incorrect line.

Subsequent research variations illustrated that the "strength" of popular opinion, or the number of confederates who gave incorrect responses, had a significant impact on whether or not the research participant gave an incorrect answer. For example, if there was even one other person who gave the correct answer, the research participant almost always also gave the correct answer. In spite of this, Asch expressed concerns that his findings seemed to highlight a "tendency to conformity in our society [that] is so strong that reasonably intelligent and well-meaning young people are willing to call black white" (1955: 5).

Asch's line experiments illustrate the question that is at the core of spiral of silence theory: how does our fundamental desire to "fit in" affect our willingness to speak out? The spiral of silence theory, developed by Elisabeth Noelle-Neumann, proposes that because people are social creatures and because we don't want to alienate ourselves, we have what she termed a "quasi-statistical organ" to help us know what others are thinking. In their overview of the first twenty-five years of spiral of silence theory research, Dietram Scheufele and Patricia Moy note:

> As a result of fear of isolation [we] constantly monitor [our] environment to check on the distribution of opinions as well as the future

> trend of opinion. Such monitoring can involve attending to media coverage of an issue, direct observation of one's environment or interpersonal discussion of issues. (2000: 9)

Television (and other mass media), therefore, help create "public opinion" on topics — especially contentious topics — such that we are able to assess public sentiment. The theory proposes that once we have a sense of what the public's opinion is, we tend to share our opinion with others (especially strangers) only if our opinion matches that of (what we believe to be) the public's opinion; if we have an opinion that is not the same, we remain silent. The spiral part comes from the fact that when minority voices are silent, the sense of a unified public opinion is strengthened and minority voices are, thus, that much more likely to remain silent. In order to test spiral of silence theory, Noelle-Neumann and other researchers have set up scenarios in which research participants actually encounter another person, or are asked to imagine encountering someone, who either shares, or doesn't, their view on a controversial topic. Based on this scenario, the participants are asked whether they would share their opinion on the same controversial topic with the stranger. This research approach is sometimes referred to as a "train test" because the context for encountering the stranger is often a mode of transportation, like a train (or plane or bus).

In general, the spiral of silence researchers find that when research participants believe that their view on a controversial topic (spanking, abortion, gay marriage) is the same as their perception of the public's opinion, then they are likely to share their view. However, if the participants believe that their view is in the minority, they remain silent. There are exceptions, and spiral of silence theory calls those who speak out regardless of their perception of the public's opinion "hard cores" and "avant gardes." Statistics for what percent of the population falls into these outspoken categories do not exist. Indeed, Scheufele and Moy label this a "neglected" area of spiral of silence research and call for research that goes beyond treating "the concepts of 'hard cores' or 'avant gardes' as assumptions rather than variables" (2000: 21). That said, while the theory continues to be refined, studies over the past thirty-five years clearly indicate a relationship between what we perceive as the public's opinion and the opinions we are willing to express in public.

One way of linking spiral of silence theory to dependency theory and agenda-setting theory is by highlighting that all three theories speak to humanity's social nature. Because we are social creatures, we rely on each other for the fundamentals of survival and joys of companionship. Dependency theory focuses on the relationship between our reliance upon mediated information and the impact that information has on our lives. Where we once gathered around the fire and depended on our elders to tell us the stories we needed to know, we now gather around television's glow and depend on the stories we receive. Agenda-setting theory proposes that mass mediated communication not only provides us with a ranking of what's important to think about but also frames the details of what to think. Similar to dependency theory, agenda-setting theory highlights that we are especially vulnerable to mediated messages about that which we have not experienced. Finally, spiral of silence theory proposes that we use mediated information to assess the public's opinion, and because we do not want to alienate ourselves from others, we tend to speak when our values match what we perceive to be the public's opinion.

Key Critical Thinkers

Social learning theory, the dependency model, agenda-setting theory and spiral of silence theory are all well-regarded theories that have been tested over the years. As a result of those tests, the theories have added to our understanding that television plays an essential role in shaping perceptions of who we are and the world in which we live. Researchers using these theories have also reflected on the implications for television's societal influence. That said, societal critique has tended to be subtle, rather than shouted, by these theories and those who research them.

Critical communication theorists, on the other hand, are loud in their critiques of society. On the quantitative-qualitative research continuum (discussed above), critical-qualitative communication theorists tend to sit on the opposite end from those who quantitatively test theories like social learning, dependency, agenda-setting and spiral of silence. For critical theorists, analyzing the role of mediated communication in creating and maintaining unjust elements of society is not an interesting concluding thought, but rather the *reason* for researching mediated communication.

Antonio Gramsci was not a communication scholar *per se*, but he was certainly one of our great political thinkers. Gramsci (1891–1937) was also leader of the Italian Communist Party. In his relatively brief life, Gramsci thought profoundly and wrote prolifically. The concept for which he is perhaps best known is hegemony. According to *Merriam-Webster*, the definition of hegemony is: "1: preponderant influence or authority over others: DOMINATION. 2: the social, cultural, ideological, or economic influence exerted by a dominant group." Gramsci's hegemony includes these elements of influence, authority over others and domination, but adds another radical element: consent from those being ruled. Gramsci's hegemony is about subjugation from above (politicians, business leaders) and consent from below (the people). My master's advisor, critical scholar dian marino, introduced me to Gramsci's hegemony. She highlights in *Wild Garden: Art, Education and the Culture of Resistance*: "Gramsci develops the notion of hegemony as the process whereby public consensus about social reality is created by the dominant class; whereby those in power persuade the majority to consent to decisions that are disempowering or not benefiting the majority" (marino 1997: 105).

At the heart of Gramsci's writings, including his thoughts on hegemony, was the Italian people's failed attempts at revolution and their support of a fascist government led by Benito Mussolini. Why would people have supported a leader who was, it seemed to Gramsci (and is now borne out by history), bound to treat them so poorly? Why would people consent to their own subjugation? Because, Gramsci argued, we come to believe that our consenting to those in power, and our sharing of their understanding of the way the world should work, is "natural." Sure a few people wield the power, have the money and exert their will on everyone else — but isn't that just the way it goes? Walter Adamson, in "Gramsci's Interpretation of Fascism," explains:

> Hegemony in this sense is nothing less than the conscious or unconscious diffusion of the philosophical outlook of a dominant class in the customs, habits, ideological structures, political and social institutions, and even the everyday "common sense" of a particular society. (1980: 627)

The masses, if you will, come to understand the world as those in power understand the world, and within that context, for example, electing Mussolini made sense. In fact it was "common sense." Todd Gitlin highlights the power of common sense in his book *The Whole World Is Watching: Mass Media in the Making and Unmaking of the New Left* when he comments that hegemonic ways of thinking come to seem "natural" to us — "as natural as living, loving, playing, believing, knowing, even rebelling" (2003: 10).

Gramsci proposed that various social institutions — schools, churches, the media — play a role in creating this climate of common sense. Of course, during his lifetime, Gramsci's notion of mediated communication was limited, essentially, to print. One can imagine, however, that Gramsci would have been awestruck by the hegemonic potential of subsequent mediated communication forms. In Gramsci's absence, other scholars have explored this potential.

Gitlin used the student uprisings of the 1960s as an example of the way in which the media play a hegemonic role. He explored how the uprisings were diffused and subsumed by the media's framing of the news related to those events. The media, therefore, not only define the world for us, but they also define our *rebellion against that world*. It therefore becomes less important what is "good" and "right" and has meaning for student protesters (Gitlin's example), and more important how the media interpret and present those meanings to the rest of the world — and back to the protesters. "As the mass media have suffused social life, they have become crucial fields for the definition of social meaning — partially contested zones in which the hegemonic ideology meets its partial challenges and then adapts" (2003: 292).

One of the most influential critical thinkers in this discussion of the role of the mass media in shaping our social reality is the cultural studies scholar Stuart Hall. What Hall's work and analysis help us to do is link television (and other mass media) to this concept of hegemony. One of the roles of television, Hall proposes, is to help us define our world such that the interests of the few are served at the expense of the interests of the many. How so? Hall says that the mass media help us to define common understandings of language.

Television plays a role in defining important concepts like "success" and "happiness." When these concepts are defined in ways that we come to believe and embrace, but that do not serve our interests, hegemony is at work. For example, think of the way in which television bombards us with stories about the importance of money and material goods in what it means to be successful and happy. When we endlessly strive to have more money and more "stuff," but wind up in debt and with little time for friends and family, our interests are not well served. Television is powerful in this process because television's stories are able to shape these concepts. (In Chapter 3 I explore television as it intersects materialism in much more detail.)

Hall articulates the power of the media in the process of shaping our understanding of the world as follows:

> Media['s] practices, among other things, represent topics, represent types of people, represent events, represent situations; what we're talking about is the fact that in the notion of representations is the idea of *giving meaning*. So the representation is the way in which meaning is somehow given to the things which are depicted through the image... on screens or the words on a page which *stand for* what we're talking about.... We're talking about the fact that it has no fixed meaning, no real meaning in the obvious sense, *until* it has been represented. (1997)

It is therefore through *representation* — such as television's representation — that so many people, events and concepts, come into being and take form for us. Who is Barack Obama? What does it look like when one is at the Oscars? What makes people happy? Television tells us. We have our own experiences of people, events and concepts, of course, and these can be distinct from television's depictions, but increasingly (as several of the theories we explored above highlight) we are reliant on media such as television to tell us about our world and our role in it. Even when we do experience something "first hand," those experiences cannot help but become enmeshed with what we receive via television.

Neil Postman's critique of television, *Amusing Ourselves to Death* (first published in 1985), shares Hall's fundamental respect for the power of mass mediated communication and also shares certain similarities with Hall's and Gramsci's notion of hegemony. Like Gramsci and Hall, Postman sees the real power of the electronic media in the fact that "the masses" readily consent to what they receive via these sources. Postman echoes Hall's and Gramci's notion of the media creating "common sense" when he argues: "There is no more disturbing consequence of the electronic and graphic revolution than this: that the world as given to us through television seems natural, not bizarre" (1986: 79).

Growing up, I knew I was affected by television. If allowed, I would consume it for hours on end (I wasn't usually allowed), and I was groggy and somewhat disoriented after viewing. The graphic and often violent images and stories would remain with me. Coming across Jerry Mander's *Four Arguments for the Elimination of Television* was a kind of revelation. I devoured his well-reasoned critiques and celebrated the audacity with which he challenged us to imagine a world without, gasp, television. He gave thoughtful articulation to the wary relationship I had developed with television. "Imagining a world free of television, I can envision only beneficial effects," Mander offered in his concluding remarks. "What is lost because we can no longer flip a switch for instant 'entertainment' will be more than offset by human contact, enlivened minds and resurgence of personal investigation and activation" (1978: 356).

As my interest in television evolved from the personal to the academic, I quickly saw how readily those in the academic community dismissed critics like Jerry Mander. More generally I was struck by the fact that television research that did not fall into the "quantifiable theory testing" category, and therefore did not appear in the best blind peer-reviewed journals, could be dismissed with a single word: anecdotal.

This is not to say that Antonio Gramsci, Stuart Hall or Neil Postman could be called lacking in "serious ideas" or that their work should even be lumped together with that of Jerry Mander. But there is still a certain "show me the numbers" demand that can be heard amongst "serious" television effects scholars. So, with this in mind, I have saved what I think is the most interesting, persuasive and exciting television effects theory for last: cultivation theory. Cultivation theory is simple in what it proposes, straightforward in how it is tested and unique in its combining of quantitatively persuasive findings and critically profound conclusions. Indeed, it is a theory that follows perfectly in the footsteps of the critical thinkers I explore here because it looks at how television encourages us to understand the world in ways that serve the interests of the powerful much better than serving our own interests. This happens not because of some nefarious plot but because it's just the way television operates. As James Shanahan and Michael Morgan propose, television creates "propaganda without propagandists" (1999: 19). It's not the *Wizard of Oz*. There is nobody *trying* to manipulate and propagandize; there is just television going about business as usual.

Bridging Quantitative and Critical: Cultivation Theory

Cultivation theory is arguably the most researched of all communication theories. George Gerbner's purpose in developing cultivation theory was to create a way to study television and think about its effects that would move beyond short-term cause-and-effect research and allow for the exploration of long-term television exposure. Therefore, rather than *narrowing* the television experience in the name of study (as is often done in quantifiable research), Gerbner was trying to expand the thinking about television such that researchers could study a medium that was not only a part of the fabric of our lives but was actually *creating* the fabric of our lives. More specifically, the theory proposes that if television affects the viewer, then those who watch more of it will be more affected. In particular, the theory proposes that the more television one watches, the more one's reality or worldview comes to resemble television's reality or worldview.

Thus, cultivation broadly begins with the premise (one you'll recognize) that television has become our society's storyteller and, as such, the medium has tremendous power and influence. Gerbner highlights in an overview of the theory that stories are hugely important in our lives because they "reveal how things work… describe what things are… tell us what to do about them" (Shanahan and Morgan 1999: ix). The actual testing of the theory, its findings and implications for the findings are, however, not simple and straightforward. This is a theory that not only *could* fill books but *has* filled books. In what follows I offer highlights of the most important elements starting with on overview of each of the three components: the "message analysis" or "content analysis" component, the "cultivation analysis" component and the "message system analysis" component.

The message or content analysis component of cultivation theory explores the themes and details that comprise television's stories. A content analysis therefore analyzes a segment of television based on elements such as the locations in which the stories are told, the props that adorn the locations, what the people look like, what the people say, what the people do and why, who the "good guys" are, who are the "bad guys" are, etc. Shanahan and Morgan state that content analysis answers the question: "What are the dominant, aggregate patterns of images, messages, facts, values and lessons expressed in media messages?" (1999: 6). This aspect of the theory also explores how television conveys "existence" (what comprises the symbolic world?), "priorities" (what is important?) and "relationships" (what/who is related and how?) (23).

Take, for example, a content analysis of violence on television — something that has been done by cultivation researchers many, many times over the years. The basic question they have asked is: "How much violence is there on television?" (Although certainly researchers have also looked at who commits the crimes on television, who solves the crimes, etc.). Most of us could probably offer an anecdotal answer to this question, but what if you couldn't just state your personal opinion about violence on television but had to *measure* the quantity of violence? What would you do?

There are almost endless ways that one could undertake the measuring of violent content on television, and there are many different television violence content analyses in the literature. Perhaps you are thinking that the task of measuring television's violence would be fairly easy because there is so much overt violence (people shooting guns, knifing, punching, kicking, etc.). Each time there is a gunshot or a stabbing or a punch, just add another instance to the tally, right? But keep in mind that there are all kinds of less overt examples of violence one would need to consider as well. What about violent/aggressive language (e.g., threatening to kill/punch/kick someone)? What about holding a gun, or pointing a gun, but not actually using it? What about verbal abuse that does not culminate in actual violence? What about "funny violence" — the classic anvil-falling-from-the-sky-and-squashing-the-coyote type violence — that takes place in a comedy or cartoon?

There are other questions you would need to consider as you set up your violence on television content measurement scheme. For example, what to do with "length of violence"? In other words, do you measure a knifing that lasts a few seconds differently from a knifing that lasts a few minutes? Or what about "level of gore"? In other words, is someone who is shot and dies without any sign of blood measured differently than someone who is shot and whose blood spurts everywhere? You get the idea. While perhaps an easy task at the surface, the measuring of content often becomes a difficult and convoluted, not to mention contentious, task in the implementation. The trick, therefore, at least in the case of cultivation research, has been to come up with a formula such that anyone who applied the same formula to the same content would arrive at the same answer about the violence in television's "symbolic universe."

One example of a formula for violence on television, called the Violence Index (VI), was developed by Gerbner and his direct theoretical descendents — Nancy Signorielli, Michael Morgan and Larry Gross. The VI is calculated based on percentage of programs with violence (%P) plus the rate of violent acts per program (RP) plus the rate of violent acts per hour (RH) plus the percentage of characters involved in violence (%C) plus the percentage of characters involved in killing (%K). The actual formula looks like this: $VI = \%P + 2RP + 2RH + \%C + \%K$ (Shanahan and Morgan 1999: 51).

So the first part of cultivation theory is the content. The second component is the actual cultivation aspect of the theory, which proposes that heavy viewers of television are more likely to understand the real world as being like the television world than are lighter viewers of television. This aspect of the theory is researched by comparing heavy (often calculated based on four hours or more of television viewing per day) and light (less than four hours per day) television viewers' responses to questions about the real world. The expectation is that heavier viewers of television will be more likely to draw upon what they know from television to answer questions about the real world. For example, if research participants are asked to estimate the likelihood of being a victim of various violent crimes (robbery, assault, rape) and heavier television viewers, on average, estimate a greater likelihood of being a victim than lighter viewers, this is the cultivation effect. I will explore how this cultivation effect is operationalized, or measured, in a moment.

The third and final component of cultivation theory is the "message system analysis." The message system is the institutional analysis of the context out of which television's messages are created and answers the question, "What are the processes, pressures, and constraints that influence and underlie the production of television content?" (Shanahan and Morgan 1999: 6). Cultivation theorists are critical of the context out of which television comes to us and are also critical of those whose interests are served by television's stories. Cultivation theorists have referred to critical social theorists like Gramsci in making the point that their real interest in exploring the implications for television's role as storyteller is to provide critical commentary about television's power.

Cultivation theorists do not believe that television is powerful because a few people are conspiring to create propaganda. They do not believe that television is powerful because people are stupid and gullible. Cultivation theorists believe that television is powerful because a small number of people tell huge numbers of people particular stories in the name of selling audiences to advertisers. This small group of people who own and create television (as researchers, such as Robert McChesney — author of *The Political Economy of Media: Enduring Issues, Emerging Dilemmas* (2008) — point out) have a narrow set of interests: present televised stories so that the largest number of people, with the largest quantities of money to spend, will watch the cheapest programming. What cultivation theorists highlight is the huge power that rests in the hands of these few people as they decide what stories to tell.

Shanahan and Morgan write: "Both [Gramsci's] hegemony and Cultivation are theories of social management which... do not assume that people are 'duped' by cultural messages, and which do not presuppose 'false consciousness.' Like hegemony, Cultivation is a theory of the power of culture over large social aggregates at the macro level" (1999: 40).

The combination of these three elements that comprise cultivation theory

— content analysis, cultivation analysis and message system analysis — is what I find so appealing about the theory. The theory offers a quantitative foundation on which one builds an unabashedly critical analysis. While other television theories and theorists have offered versions of this quantitative research/critical combination, no theory has combined these elements as successfully, or made as profound a contribution to our understanding of what television is doing to us and the world in which we live, as has cultivation theory.

Starting with Violence

One of the most compelling, enduring — and grappled with — stories that television tells is that of violence. No area of cultivation research has garnered more attention than violence. George Gerbner and his colleagues began their Cultural Indicators project in the early 1970s in order to monitor the content of violent programming. The project has provided the researchers with a detailed longitudinal tracking of television's violent content (tracking that continues to this day). Their content analyses uncovered such things as the rate at which violence occurs in television's world and the proportion of the television population that is involved with violence.

For example, in an early cultivation study, undertaken 1976, Gerbner and Gross found that while close to two-thirds of people in fictional television programming were somehow involved with crime, the FBI statistics showed the real world number to be less than half of 1 percent. In the same study, the researchers found that while the U.S. census indicated that 1 percent of employed men in the United States made a living in law enforcement, closer to 12 percent of the U.S. population was shown in such an occupation on television. With this knowledge of television's content, the researchers could then move on to the actual cultivation aspect of the theory and compare heavy viewers' (four hours plus of television viewing, on average, per day) and light viewers' (less than four hours of television, on average, per day) responses to questions about violence in the real world.

Research participants might be given a list of questions about violence, such as: "What is the percent of people who walk through Central Park in New York City who are attacked?" "What is the percent of women who are raped in their lifetime?" "What percent of people have their homes broken into at some point in their lifetimes?" The point in asking these questions is not to see whether participants are "right" or "wrong" in their answers, or even how accurate people's responses are. Instead, the questions allow researchers to compare light television viewers' estimates with heavy television viewers' estimates. In general, people are not accurate in their assessments of frequency and levels of societal violence (the real world violence percentages are very small, and almost without exception everyone guesses incorrectly), but what the cultivation research consistently finds is that estimates by heavy television viewers are *higher* than those by light television viewers.

Therefore, what cultivation of violence research has uncovered over the years is that heavier viewers of television are more likely to answer questions about violence in the real world with responses that resemble the violence in television's world. It would seem, therefore, that instead of drawing on their own experiences, heavy television viewers draw upon their television experiences. In the context of violence, this means that heavier television viewers think of the world as a more violent place than do lighter viewers. Cultivation researchers call this the "mean world syndrome" and have found this "complex of outlooks which includes an exaggerated sense of victimization, gloom, apprehension, insecurity, anxiety and mistrust" (Shanahan and Morgan 1999: 55) again and again.

Television's stories of violence are, however, only one category of the stories that television tells. Cultivation researchers have also looked at the relationship between levels of television viewing and attitudes about, and understandings of, politics, sex roles, race, body image, age, gays and lesbians, materialism and the environment (see Shanahan and Morgan 1999 for an overview). These various areas of cultivation research represent thousands and thousands of studies. In an attempt to make sense of whether there was some kind of "average finding" in cultivation studies over the years, Morgan and Shanahan undertook a meta-analysis of twenty years of research. A meta-analysis essentially looks at a number of existing studies and turns each study's findings into data that can be averaged. Morgan and Shanahan (1997) took eighty-two cultivation studies, comprising 5,633 findings, and calculated an average effect size. The researchers found that while television's contribution to people's understanding of the world may be small (r = .091), it is consistent. And as Shanahan and Morgan assert, Gerbner was cautious about television's influence, pointing out that his intention was "not to assert that television alone is responsible or necessarily decisive, only that it makes a contribution" (cited in Shanahan and Morgan, 1999: 57). But almost twenty-five years later, Shanahan and Morgan were less reserved when they asserted: "Television is by no means the most powerful influence on people, but it is the most common, the most pervasive and the most widely shared" (33).

Cultivation researchers therefore highlight television's pervasive, widely shared contribution and point out that we live in a world that increasingly respects the importance of small changes that occur on large scales. Take, for example, Shanahan and Morgan's climate change analogy (1999). While climate change proposes that average global temperatures will change by only a few degrees over the next number of years, the entire global climatic balance is being thrown off because of these small changes — and the implications are massive. What cultivation research highlights is that while short-term cause-and-effect television research can be interesting, there is very real and persuasive evidence that television has been impacting our understanding of the world in subtle, yet profound and ongoing ways since its arrival in our lives. The impact of television should be causing us to take real and serious notice.

Psychology of the Cultivation Effect

The establishment of the "reality" of a cultivation effect led naturally to the question of *why* there is a cultivation effect. People have so many sources of information about the world around them, how is it possible that one single information source, television, can make enough of a difference that time and again researchers have been able to show differences between heavy and light viewers? It is an interesting question but one that cultivation researchers have not, in general, felt particularly compelled to try and answer. As Shanahan and Morgan point out, cultivation theory was developed to

> examine "the bucket not the drops" of television's contribution to the culture. Indeed, we believe that to become sidetracked by the peculiarities of how individuals receive, process, interpret, remember and act on messages can distract attention from the more central questions of cultivation research. (1999: 172)

But a few researchers, such as L.J. Shrum, have gone about exploring why the cultivation effect exists.

The essence of what Shrum (2007) proposes is that the difference between heavy television viewers and light television viewers has to do with heuristics, or the mental short-cuts that we use when we think about the world. We use heuristics because we are "cognitive misers" (i.e., we only have so much mental space and energy, so we conserve them). What this means is that heavy viewers have television's version of the world more readily available cognitively than light viewers, and thus, when asked about violence or sex roles or politics or anything else in the "real world," heavy television viewers draw upon what's right there: television's reality.

One way that Shrum has highlighted this cognitive phenomenon has been to manipulate how much people are able to ponder and process what they have been asked. Shrum has found that the more thoughtful people are able to be in their answers to questions about violence, for example, the smaller the gap between heavy viewers' answers and light viewers' answers. How can processing time and thoughtfulness be manipulated? In a study published in 2007, Shrum had two groups answer exactly the same survey: one group was sent the survey via "snail mail" and participants answered the questions in their home at their leisure; the other group answered the same questions on the phone while someone waited for the answers.

As predicted, the gap between heavy and light viewers' responses was greater for the on-the-phone-group because they had less time to process the answers (i.e., they felt rushed because someone was waiting for their answers). The heavier viewers were, therefore, more likely to draw upon what was most available, their "television answers." When people answered the survey at home, however, people were able to thoughtfully rule out what they knew from television versus what they know from their "actual lives," and thus the cultivation effect was diminished.

Another way that the effect of cognitive processing is seen in cultivation research is when questions about television viewing *precede* the other questions on a survey; for example, if questions about one's television habits precede questions about violence, then there is no cultivation effect. The assumption is that the cultivation effect is eliminated because people have been alerted to television's role by asking them about their viewing habits, or in psychological terminology, the participants were "primed" by the television questions to think about television. Once people are thinking about television, they are able to filter out the information they have received from television and answer the questions based on their lived reality.

Another aspect of cultivation research is that it tends to be concerned with viewers' thoughts about the world, not their behaviours. In other words, while cultivation research has shown conclusively, in a large number of contexts, that there are differences between how heavy viewers and light viewers *think* about the world, the research does not make claims about differences between how they *behave* in the world. Not surprisingly, there is plenty of research that shows that there are relationships between thoughts/attitudes/values/etc. and behaviours, but these relationships are surprisingly complex. So, cultivation theory does not make behavioral claims (i.e., just because a heavy viewer thinks the world is a meaner and scarier place does not mean that the viewer behaves in ways that reflect those thoughts). That said, often researchers are able to link cultivation findings to research that illustrates how attitudes and behaviours are connected.

Cultivation Theory's Critics

The biggest targets are the easiest to hit. All of the theories and theorists I have explored in this chapter have their critics, none more so than cultivation theory. Cultivation theory researchers have produced a tremendous amount of research over the years and have made many claims, giving potential critics a big target, indeed. Not only have the results of cultivation research been held up for critique, but so too has the methodology. Take, for example, the first step of almost all cultivation studies: a content analysis. The content analysis allows researchers to claim that they know what television's version of violence (or sex roles, or the environment, etc.) is, but a critic can always say "that's not the way *I* would define violence." This critique has been fairly readily addressed because there does not need to be an "ultimate" definition of any of these concepts, only a definition that reasonably taps into what happens on television such that heavy and light viewers can be compared.

Critics of cultivation theory have also latched onto the phenomenon that by controlling for some demographic variables, like gender and where a person lives, the cultivation effect can be reduced or eliminated. In other words, the critics said that the sweeping effects that cultivation researchers were claiming, such as heavier television viewers seeing the world as a meaner and scarier place, wasn't necessarily what was going on. When cultivation researchers looked into this critique, what they discovered allowed them to develop the theory rather than diminish it. The two key concepts that came out of this critique — that the controlling of demographic variables erased the cultivation effect — are known as "mainstreaming" and "resonance."

Mainstreaming proposes that while people are affected by television based on the quantity of television they view, people are also affected by television's content as it relates to their experiences of the real world. For example, when light viewers of television are asked to reflect on how mean and scary a place New York City is, women are likely to provide higher estimates of the dangers of the city than are men. This is presumably because women are more likely to have grown up understanding that they can be vulnerable to various forms of violence, like rape, and they are more likely to have been taught to actively avoid such threats by not traveling alone, staying in well lit areas, perhaps taking a self-defence course, etc. Men, on the other hand, are less likely to have grown up being taught about the violence that lurks in parks, after dark, etc. This isn't to say that men don't experience violence (in fact, men can be more likely than women to experience certain types of violence), but in general boys do not grow up being taught to be wary of violence in the same way, or with the same intensity, that girls are.

So, there is a gap for light television viewing men and light television viewing women, with women understanding their reality as meaner and scarier. Here's the mainstreaming part: when men and women are heavy viewers of television, both genders experience a cultivation effect such that men's and women's scores go up (versus their lighter viewing counterparts) on their assessment of the world as a mean and scary place — but men's scores go up more. Why? Because men's experience of violence in the *real world* tells them it's safer for them than it is for women, but men's experience of violence in the *television world* tells them the world is *just as violent* for them as it is for women (and television's world is very violent!).

Some cultivation researchers have likened this mainstreaming effect to television's gravitational pull, where "the angle and direction of the 'pull' depends on where groups of viewers and their styles of life are with reference to the center of gravity, the 'mainstream' of the world of television" (Shanahan and Morgan 1999: 73). In other words, television has a particular reality into which heavy viewers are constantly being drawn. This gravitational pull works on all viewers. But for viewers whose lived reality is that much "further," or different, from television's, television's pull is that much stronger.

Resonance is another concept that researchers developed in response to criticism that in some cases controlling for differences in where one lives can eliminate the cultivation effect. For example, some research demonstrated that there was a "mean world syndrome" in urban areas that really did have more violence than other areas. In other words, perhaps heavier viewers of television in these urban areas thought the world was meaner and scarier because where they lived *really was meaner and scarier* (and perhaps they were watching more television because they were afraid to go outside!).

With further investigation, researchers found that rather than discounting the cultivation effect, the phenomenon of television's violence mirroring some people's real life experiences of violence was causing "resonance" — or an *amplification of the cultivation effect*. In these situations, therefore, television confirms, even affirms, and highlights the very real violence in people's lives. Light television viewers from a violent part of an urban area might have higher cultivation scores than heavy television viewers from a less violent area, but when heavy and light viewers *from the same violent area* are compared, heavier viewers score higher than lighter viewers.

Here's a final thought about "what comes first?" and why we can assume that television does, indeed, teach us about things like violence (and it isn't that those who already think of the world as mean and violent just watch more television). Think back to Shrum's research into the cognition of the cultivation effect — he can help us to establish causation. His research shows that when people are asked about television before the other questions or people are simply given enough time to thoughtfully process their answers, then the cultivation effect is decreased or eliminated altogether. If people were turning to television because of pre-existing understandings of the world, then highlighting television's role in our lives, or giving more time to think about television's role in our lives, would not affect the responses.

One final critique of cultivation theory relates to the question: "Doesn't it matter *what* we watch?" Some cultivation research has focused on the effects of specific types of programming (and in the next two chapters I highlight a couple of examples). The real answer to the content question, however, is that while the effects of individual messages and genres of programming are interesting to explore, it is not what cultivation explores. Cultivation researchers investigate story systems more generally — the context out of which television's stories are born, the ingredients of television's stories and the effects of those stories.

Don't get me wrong. We are all unique individuals, and we all come to watch television as unique individuals with unique tastes and interests. But what the cultivation research highlights is that as we watch television there's a certain "melting pot" phenomenon. The more we watch television, the more our worldview becomes like television's worldview and like the worldview of other people who watch a lot of television.

The Conclusion? We Are Affected

While the various theories about and ways of looking at our relationship with television can be very different from one another (and even highly critical of one another), it's important to highlight the similarities. Every theory discussed above, and I would argue that every theory that has ever explored television, has as a basic "finding" that there is power, or influence, in television's storytelling role. The theories and theorists focus on different aspects of television's power and influence — to socialize us like other socializing factors (social learning theory), to tell us about the world when we don't have a personal context for the stories (dependency theory), to tell us what's important (agenda-setting theory), to tell us what to think (framing), to create "public opinion" and to silence us (spiral of silence theory), to instill a "common sense" understanding of the world (Hall) that can cause us to be willingly subjugated (Gramsci), to turn everything into amusement (Postman), to cause us to draw on television's world when thinking about the real world (cultivation theory) — but they all concur that *television affects us*.

As the reactions to some of the early books that were highly critical of television — Mander's *Four Arguments for the Elimination of Television*, for example — can attest, it used to be that speaking out about the dangers of television often garnered negative attention. Perhaps this was because of television's homey, cozy beginnings that involved families gathering in the living room to view together; perhaps because of television's seemingly benign role as entertainer; or perhaps because, as uses and gratifications research points out, we *like* to sit and watch television.

Concerns about the links between television and violence led the way for "acceptable," even desirable, critical thinking about television. The arrival of a rating system and the rating system's warnings about television's potentially problematic content (e.g., announcements before certain shows that the content may not be appropriate for some audience members) and the development of the v-chip (to allow for parental blocking of content) made explicit the television industry's acceptance that some concerns are warranted. That said, I think that one of the most striking examples of concern about television and its effects are the warnings from the American

Academy of Pediatrics, the Canadian Pediatric Society and the American Medical Association.

The Canadian Pediatric Society's statement, *Impact of Media Use on Children and Youth*, originally came out in 2003 and was reaffirmed in 2011. The American Academy of Pediatrics statement, *Children, Adolescents and Television*, was published in 2001. Both documents, which draw upon extensive research, offer cautionary tales about young people's relationship with television. Not surprisingly, concern about violence is present, but it is only one concern of many. Also discussed are time displacement (that television reduces time for reading, exercising, playing with friends); nutrition and obesity (that television has been linked to unhealthy eating habits, sedentary lifestyles and obesity); poor learning and academic performance; sexuality (inappropriate and prolific images and information about sexuality and links to irresponsible sexual behaviour); alcohol, smoking and drug use (passive promotion of these products and evidence that such exposure can lead to early use); and poor body concept and self-image. The *Journal of the American Medical Association*'s article "Television Viewing and Risk of Type 2 Diabetes, Cardiovascular Disease, and All-Cause Mortality," published in June 2011 (Grontved and Hu) sends a clear and cautionary tale of the kinds of risks people around the world take as they choose to spends hours of each day watching television.

So, yes, the grand conclusion is that television adversely affects us, and the more we watch, the worse it gets. Even these arguably conservative and certainly scientifically savvy medical associations have looked at the television effects research and have come to this conclusion. At some point as I make my way to this conclusion with students, a question arises that goes like this: But if television has the ability to adversely affect us, doesn't it also have the ability to positively affect us? What if television told *different* stories? What if it told *better* stories?

Sesame Street, the famous children's show so many of us grew up with, is often held up in this context of television's capacity to be a positive, or "prosocial," force. And, indeed, as the Canadian Pediatric Society's statement on media use offers: "The educational value of *Sesame Street* has been shown to improve the reading and learning skills of its viewers. In some disadvantaged settings, healthy television habits may actually be a beneficial teaching tool." The American Academy of Pediatric's statement on television similarly offers a few positive words: "Although there are some potential benefits from viewing some television shows, such as the promotion of positive aspects of social behavior (e.g., sharing, manners, and cooperation)," but finishes the sentence with, "many negative health effects also can result" (2001: 423). And that's it. In both cases, therefore, the concerns these associations raise about television's problematic reality far outweigh any promise a few shows have offered for television's positive potential.

So the answer to the question of whether television can tell "better" and "different" stories is yes. It can. *Sesame Street* does. But *Sesame Street* and similar programming represent but a few drops in television's ocean. Television is best understood as an ocean with waters that, ultimately, have nothing to do with pro-social or anti-social messages; television's ocean is about selling. As Mander notes: "To have only businessmen in charge of the most powerful mind-implanting instrument in history naturally creates a boundary to what is selected for dissemination" (1978: 264). That's the crux. To get bogged down in television's theoretical potential "for good" is not worth the energy. Television's reality is that it is both our society's storyteller and a massive business venture with one goal: selling us more and more stuff. This is the most fundamentally important and devastating aspect of television. Television's primary role is that of promoting materialism. Yet this role has lurked in the shadows and been largely overlooked. In the next chapter I turn on the spotlight.

3.   MATERIALISM

In this chapter I build on the Television chapter's foundations as I explore a woefully under-represented aspect of television research: how television affects our relationship with the material world. In fact, my goal in this chapter is to convince you that television's relationship with materialism is not just fundamentally important, but fundamentally devastating to us and to the planet (but I'll save the planetary piece for the next chapter). For now I focus on materialism —the important ways in which television and materialism are connected, and why we should care.

Meaning of Materialism

How do we relate to and interact with the stuff in our lives? This is a question that sits at the heart of materialism. Or, as Russell Belk states, materialism is "the importance a consumer attaches to worldly possessions. At the highest levels of materialism, such possessions assume a central place in a person's life and are believed to provide the greatest sources of satisfaction and dissatisfaction" (1985: 265). This definition of materialism revolves around the importance people attribute to their possessions. The more materialistic people are, the more likely they are to believe, for example, that their possessions are what define them.

Marsha Richins and Scott Dawson's (1992) eighteen-statement scale measures materialism. You might find it interesting to calculate your score. Respond to the following statements where 1=strongly disagree; 2=disagree; 3=not sure; 4=agree; 5=strongly agree:

1. I admire people who own expensive homes, cars and clothes.
2. Some of the most important achievements in life include acquiring material possessions.
3. I don't place much emphasis on the amount of material objects people own as a sign of success.
4. I usually buy only the things I need.
5. I enjoy spending money on things that aren't practical.
6. I try to keep my life simple, as far as possessions are concerned.
7. I have all the things I really need to enjoy life.
8. My life would be better if I owned certain things I don't have.
9. The things I own say a lot about how well I'm doing in life.
10. I like to own things that impress people.
11. I don't pay much attention to the material objects other people own.

12. The things I own aren't all that important to me.
13. Buying things gives me a lot of pleasure.
14. I like a lot of luxury in my life.
15. I put less emphasis on material things than most people I know.
16. I wouldn't be any happier if I owned nicer things.
17. I'd be happier if I could afford to buy more things.
18. It sometimes bothers me quite a bit that I can't afford to buy all the things I'd like.

As is often the case with research scales, some of the statements on the above scale go in the "opposite direction" of other statements. This is done to try and identify any people who aren't paying attention to what they're doing and are providing the same response for every statement or providing incongruent responses. Therefore, when you add up your scores, you'll have to switch around the "reverse-coded" statements and make sure that all of your calculations go in the "same direction." A higher score is indicative of greater materialism.

It is also important to note that while I want to explore the above definition of materialism, the concept of "materialism" as an area of study is often much more esoteric than "we are what we own." For example, John Smart, in his *Journal of Philosophy* article entitled simply "Materialism" offers: "By 'materialism' I mean the theory that there is nothing in the world over and above those entities which are postulated by physics (or, of course, those entities which will be postulated by future and more adequate physical theories)" (1963: 651). According to Smart, therefore, what is covered by the concept of "materialism" is what is contained in the physical world. On the other hand, quantum physicists disagree and propose that the world is much less about solid matter than we may believe. Such philosophical discussions are well beyond the scope of what we're exploring here.

You may be wondering what the difference is between "materialism" and "consumerism." The online Merriam-Webster dictionary defines consumerism as "the theory that an increasing consumption of goods is economically desirable" and "a preoccupation with and an inclination toward the buying of consumer goods." So while materialism and consumerism share the drive to buy and consume, materialism includes the *reason* for that consumption; materialism links what we own with who we are. That said, materialism and consumerism are often defined in ways that make them interchangeable and indistinguishable. For example, economist Paul Ekins defines consumerism as the "possession and use of an increasing number and variety of goods and services [in the belief that this is the] *surest… route to personal happiness, social status, and national success*" (1991: 243, italics added). Therefore, while acknowledging that there can be overlap in the definitions,

I use the term materialism exclusively in what follows with the understanding that it includes both the desire to consume and a sense of being defined by that which is consumed.

Newness of Our Materialism

In some ways, materialism has always been a part of the human experience. In other ways, the materialism that so many of us currently possess and the materialism which the Richins and Dawson scale addresses is brand new. In the 2010 article "Butchering Dinner 3.4 Million Years Ago," from the journal *Nature,* Richard Lovett highlights skeletal evidence indicating that 3.4 million years ago stone tools were being used to butcher animals. Very early humans therefore had material possessions which were fundamentally important to their lives. Less functional items arrived thousands of years ago. According to anthropologist Dennis O'Neil (2011), the earliest musical instruments (such as bone flutes) and figurines were made about 35,000 years ago. We humans have, therefore, had possessions to help us acquire and consume food, create shelter, tell stories, celebrate important rites of passage and so on for a long time. Our ancestors imbued those material objects with great value and meaning, but the value and meaning were directly related to the value of the function that the object performed in our personal and social lives.

North Americans' current relationship with the material world is very different from these pragmatic early days, and the origins of this change in our relationship with the material world can be found a few hundred years ago (a blink of an eye!). Senior fellow at the Worldwatch Institute, Erik Assadourian, proposes that in the late 1600s:

> Societal shifts in Europe began to lay the groundwork for the emergence of [materialism]. Expanding populations and a fixed base of land, combined with a weakening of traditional sources of authority such as the church and community social structures, meant that a young person's customary path of social advancement — inheriting the family plot or apprenticing in a father's trade — could no longer be taken for granted. People sought new avenues for identity and self-fulfillment and the acquisition and use of goods became popular substitutes. (2010: 11)

On the horizon, the Industrial Revolution was beginning to usher in a new era in which acquisition of these "popular substitutes" would be easier than ever before.

The Industrial Revolution, commencing in Britain in the eighteenth century and arriving in the United States by the mid-nineteenth century, was a result of, and resulted in, massive technological and social shifts. During this time there were significant changes in, and interplay between, energy (the steam engine is sometimes held up as the symbol of the Industrial Revolution), transportation infrastructure, iron production, coal mining, textile manufacture and agricultural practices. The population grew as death rates dropped (the end of various plagues helped with this), birth rates increased and food was generally more plentiful. People flowed into cities, bringing about the urbanization that was necessary for industrialization. Indeed, it took nothing short of a revolution to change the way we relate to the material world. What had been largely a world of small-scale subsistence hunting and gathering-based communities, and later farming-based communities, suddenly changed as steam and coal-powered factories were built. These changes meant that large numbers of people were part of manufacturing processes, most notably the assembly line, that were producing goods at an unprecedented rate.

There are many ways to encapsulate the huge number of changes that the Industrial Revolution ushered in, but this is certainly one: it allowed for the production of goods at an unprecedented rate. This new production reality brought with it a fundamental challenge. The only way that this new abundant supply of goods could make economic sense was if there were purchasers. Stuart Ewen notes in *Captains of Consciousness: Advertising and the Social Roots of the Consumer Culture*: "With a burgeoning capacity, industry now required an equivalent increase in potential consumers of its goods" (1976: 24). Thus, for the first time in history, large numbers of people were not only needed to *produce* huge quantities of goods, large numbers of people were also needed to *consume* huge quantities of goods. I offer this equation: the birth of the assembly line equals the birth of materialism — with the recognition that it hugely simplifies a complex and contentious history. In his article "Crossing Divides: Consumption and Globalization in History," Frank Trentmann cautions against simple linear models for explaining the spread of materialism. He questions the notion that the nineteenth century ushered in an era of "things and shops." He acknowledges, however, that what sets the seventeenth and eighteenth centuries apart from earlier periods is that "there was a deepening and broadening of consumption... [whereby] goods played an ever more important function as social positioning devices in increasingly complex, growing urban environments" (Trentmann 2009: 195).

Our relationship with the material world has changed quickly and dramatically since the Industrial Revolution, and this discussion of recent dramatic change may sound similar to what I explored in the Television chapter about the recent dramatic changes in how we communicate with one another. In the Television chapter, I noted that we have spent most of human history communicating face-to-face and telling stories about that which was fundamentally important to our lives — finding food and shelter and pondering where we came from and what happens when we die.

When mediated communication, especially television, arrived in our lives, it changed who told the stories and what the stories were about. Suddenly very few people were crafting stories for huge numbers of viewers with the goal not of imparting vital information about our lives, but rather with the goal of entertaining us such that the largest audiences could be amassed and sold to advertisers for the greatest profit.

Similarly, throughout most of human history we owned few material objects, and what we did own had important and functional roles in our lives. Then suddenly, and generally coinciding with the arrival of mediated communication, our relationship with the material objects in our lives changed. The stories we were being told increasingly challenged us with the notion that our relationship with the material world need not be functional and utilitarian. In fact, the stories encouraged us to understand that the material world could, indeed should, be about "social positioning." As for how much to own, the mediated stories encouraged us to understand that we could never possibly have enough possessions.

It is interesting, therefore, to highlight that these two fundamental changes — with our stories and our stuff — were set in motion at about the same time and were intertwined. The assembly line revolution enticed us to uproot ourselves from our small communities and head to the urban centres for jobs. The increasing population density, literacy and mechanization also created the perfect context in which mediated communication could flourish. In our uprooted urban lives our reliance on these new mediated stories grew, and as the unprecedented quantity of goods spilled off the assembly lines, we looked to these stories to help us make sense of how we should relate to all this stuff. "Sex, food, wealth, power, prestige: they lure us onward, make us progress," offers *A Short History of Progress* author Ronald Wright. "And to these we can add progress itself, in its modern meaning of material things getting better and better, an idea that arose with the Industrial Revolution and became its great article of faith" (2004: 81). This I believe sums up our current relationship with the material world; we are what we own (and we have to commit to consumption because "better stuff," that could improve who we are, is always being created).

Not everyone agrees. Some researchers, most notably political scientist and materialism scholar Ronald Inglehart, would add a caveat to what I propose about the history of our relationship with materialism. Inglehart believes that since the Second World War, members of more affluent parts of the world have been moving towards "*post*-materialism," or beyond our focus on what we own, and increasingly placing "emphasis on autonomy, self-expression and the quality of life" (2008: 131). Inglehart has based his findings on asking research participants the following question (also known as the Inglehart index):

There is a lot of talk these days about what this country's goals should be in the next ten or fifteen years. Would you please say which one of them you yourself consider most important in the long-run: a) Maintaining order in the nation; b) Giving people more say in important political decisions; c) Fighting rising prices; or d) Protecting freedom of speech.

Inglehart believes that when a) and c) are chosen, participants display a "materialist" or "acquisitive" orientation, whereas b) and d) reflect a "post-materialist" or "post-bourgeois" orientation. Over the almost forty years in which Inglehart has been tracking responses to his index, he has found that people in more affluent parts of the world are increasingly post-materialist. He attributes these changes to people's "existential security," or the fact that people in these parts of the world are increasingly able to take survival for granted.

I propose that the post-materialism that Inglehart is describing and the attachment to our belongings that is present in Belk's definition of materialism (and is measured by the Richins and Dawson scale) are not mutually exclusive. It makes perfect sense that as we feel secure in our lives, our interest in political decision-making and freedom of speech would increase. But just because we can expect that the nation will maintain order and just because rising prices are, relatively, less important to many of us, it doesn't mean that our love affair with "new stuff" is going away. Indeed, as you will see in what follows, our love affair with stuff is alive and well.

How We Learned Our New Materialistic Role

Advertising is often highlighted as being key in how people learn to be materialistic. Indeed, what distinguishes the earliest advertising is its educative earnestness. These early print advertisements are notably different from the advertising of today — for example, early ads often contained a lot of what was, essentially, instructional text. This text was not just an attempt to convince people of the virtues of a particular product but also often included users' product testimonials and details on how to actually *use* the product. Here's a testimonial entitled "Fresh and dainty after a long motor ride," from Grace Frisby, New York City, which accompanied a 1924 Ponds Vanishing Cream print advertisement:

We nearly always have our luncheons, dinners and dances at the country club. It's a darling place but kind of far out — a fairly longish motor ride. Most of us look like shiny-nosed sights when we get there but one particular girl always looks lovely. One night I plucked up the courage to ask how she did it and she told me about

Ponds Vanishing Cream. I use it too now. I find that it protects my skin from the dust and wind and holds the powder on smoothly for hours. I love the soft feel of it and the nice *finished* look of it. (Pond's Extract Company)

Such long-winded context and explanations made a lot of sense given that so many people had never really consumed beyond the essentials before. If one is fed and clothed with adequate and comfortable shelter, what else would one need? One would need to be told that finishing cream is important and why. The Industrial Revolution brought assembly lines that were chugging along, and those who were dutifully working on them were going to have to learn to consume faster, and in much greater quantities, if the new industrial system was going to survive.

It was clear that the historical notion of owning little beyond the essentials was not going to be good enough; neither was buying based upon practicalities (just how "practical" is "finished" skin?!). We were going to need to learn to consume far beyond utility. Ewen offers: "The utilitarian value of a product or the traditional notion of mechanical quality were no longer sufficient inducements to move merchandise at the necessary rate and volume required by mass production" (1976: 34). What appeal might provide the sufficient inducement? It's right there in the Ponds advertisement. Without more stuff, we are incomplete. "The negative condition was portrayed as social failure derived from continual public scrutiny... the positive goal emanated from one's modern decision to armor himself [or herself] against such scrutiny with the accumulated 'benefits' of industrial production" (1976: 36). Advertisers realized that the power of any given product lies not in the actual, utilitarian qualities of the product but rather resides within the self-doubts of the consumer. All the consumer needs to believe is that the very *act of purchasing* a product might alleviate some self-critical discomfort. All the advertiser needs to do is provide ongoing critique.

Advertising (and as I discuss below, television more generally) rose to the challenge. While the path to changing our consumption habits hit some bumps along the way — world wars, the Great Depression — advertising helped us to learn to be exemplary consumers such that today we do a pretty good job of keeping pace with mass production. The subtleties of how we were taught to change our consumption expectations and habits, and the complexities of how persuasion works, fill endless academic and popular books and articles. As Canadian advertiser/marketer and radio broadcaster Terry O'Reilly comments on his website, advertising and marketing "heave under reams of research."

In her cultural history of advertising research, Peggy Kreshel proposes that there are two primary reasons for the huge quantity of advertising research that has been conducted over the years. First, Kreshel suggests that the research has been used "as a scientific tool to reduce risks associated with uncertainty" (1993: 60). Advertisers and marketers wanted research to provide them with cause-and-effect formulas for how to get people to consume. For example, Stanley Resor — who purchased the advertising powerhouse J. Walter Thompson Company in 1916 — believed that in the same way the physical world responds to certain laws, so too were there laws governing human behaviour. The power of research was, therefore, to uncover these behavioural laws as they pertain to advertising and persuasion. In his 1921 book *Do Laws Really Govern?* Resor argued the following: "Whenever one of us goes to a theater, or picks a necktie, we are responding to definite laws. For every type of decision — for every sale in retail stores — basic laws govern the actions of people in great masses" (cited in Kreshel 1993: 66). Once we know the laws of persuasion, the guesswork — and risk — in advertising will be removed.

Kreshel claims there has also been such a large quantity of research about persuasion and advertising because it has been used as a "professional symbol of the legitimacy of [the] advertising practice" (1993: 60). Advertisers have felt that being associated with "serious research" makes them and their profession look good.

Not everyone in the advertising world agrees on the value of such research. Indeed, immediately after O'Reilly's comment on his website about the advertising industry heaving under reams of research, he says: "I have always believed that, at the end of the day, it [advertising/marketing] still comes down to creativity and persuasive ideas. And persuasion is an art." But the quantity of research about advertising that has been done over the years leads us to believe that while some advertisers may want their occupation to be an art form, they can't run the risk of missing out on what science can tell them. Kreshel states: "Proponents of scientific [research-based] advertising did indeed believe that science would enable them to *control the consumption habits of non-rational consumers*" (1993: 64, italics added).

According to the Communication and Mass Media Complete database (a collection of 820 communication and mass media publications — an extensive, but certainly not exhaustive, collection), there were over 73,000 articles related to advertising ("advertising" appears in the title, subject terms, keywords or abstract) written between 1915 and 2012. So, for those who are interested in what a more scientific/research-based approach might unearth, there is plenty of reading available. And while we may question whether advertising research has facilitated the "control of consumption" or found the "laws of persuasion" for which advertisers like Resor once hoped, it is certainly clear that a lot of time, energy and expense have gone into exploring a huge number of facets of advertising and marketing (and how, more generally, people can be persuaded to consume as much stuff as possible).

For years, therefore, advertisers have worked to understand how to encourage us, as a species, to change the way in which we have historically consumed. A part of this process is the way in which advertisers have worked to understand how to encourage those newest to the world, our children, to consume. One category of children's advertising research that I find interesting (and clever) is what I will call the "Santa Claus studies." These studies, starting in the early 1970s, have looked at the relationship between children's exposure to television advertising and their gift wishes as expressed in letters to Santa Claus. What this research has discovered, in laboratory and field settings (with younger and older children, and in various countries) is that the more television children watch, the more likely it is that those children will ask Santa for the specific products advertised on television. Professor of developmental psychology Karen Pine indicates:

> Our studies show that the more TV children watch, the more toys they ask for in their letters to Santa. Children age four or five ask for more toys but don't usually mention "brand" names. So they'll ask for a "baby doll" but not "Baby Annabel" by name. By the time kids are six to eight years old, though, they not only ask for the advertised toys by brand name, but also they even tell Santa which stores he can buy them from and the retail prices. (2008: 17–18)

Advertising not only teaches about the desirability of material items, however, but also about the social world in which those items exist. For example, in the late 1970s, researchers Marvin Goldberg and Gerald Gorn studied young children (four and five years old) to explore some of the implications of their exposure to advertisements. Their study involved randomly assigning the children to watch one of two versions of a ten-minute television show. In one version of the show there were two advertisements for a toy and in the other version of the show there were no advertisements. After the program was over, the children were shown two pictures of "equally attractive" (important to note) little boys. One of the boys was described as "not very nice" and he held the advertised toy. The other boy was described as "nice" and he did not hold a toy. The participants were then asked two questions: "Which child would you like to play with?" (the options being play with the not nice boy with the toy or play with the nice toy-less boy) and "Where would you like to play?" (the options being play with the toy alone, or play with friends, but not the toy, in the sandbox).

Sixty-five percent of the children who had seen the program with the advertisements wanted to play with the not nice child who had the toy (only 30 percent who had seen the program without the ads wished to play with this child). Conversely, 70 percent of the children who had seen the program without the advertisements wished to play in the sandbox with friends (only 36 percent of the children who had seen the program with the advertisements wanted to play with friends in the sand). The materialism in the advertisements, therefore, not only encouraged the viewers to covet the toy but also encouraged them to make socially undesirable choices.

Children, in their vulnerability to the stories they're told, can be seen as a microcosm, or metaphor, for all of us. Sure we may not watch advertisements and then include the advertised toys in a letter to Santa. We're too grown up for that. After all, we don't write letters to Santa anymore, and we may even do our best to avoid television's advertisements altogether (at least the explicit ones over which we might exert some control). But I propose that our relationship with the mediated messages of materialism changes throughout our life; we may be able to look back on our relationship with the advertised world of materialism and find it quaint, but that's only because the relationship has evolved, not because it has ended.

Advertising's Success Story

There are many ways to make the case that our transition to being people who are, in Ewen's words, "habituate[d]… psychically to consumption" has been successful. One way would be to look at how nuanced the art and practice of materialistic persuasion has become. Where ads were once text-heavy and instructional, they have now become vague and often whimsical in their presentation. Where advertisers once felt compelled to share with us the attributes of products and how we would actually use them, now ads imply what it would feel like to be in the "midst" of a product. We are shown what cool friends we'll have when we have this product, what our love life will contain, how happy our family will be. There's much less fact and much more fiction in advertising. Indeed, today it is not uncommon for advertisers to de-emphasize the product that is being advertised — sometimes leaving it up to the consumer to decipher what product is actually being advertised.

One way to see our radical historic shift from "material minimalists" to "material maximists," therefore, is by noting the changes that have taken place in how we are appealed to as consumers. At one time advertisers had to sell *themselves* as necessary to the industrial process. Today, advertising's success can be seen in our collective embracing of our consumer role. Indeed, advertising's endless materialistic appeals that link the very essence of who we are with that which we own and covet are as matter of fact and as unremarkable as the air we breathe.

There are however, other indicators of how our relationship with the material world has shifted dramatically in the last hundred or so years. In *Plentitude: The New Economics of True Wealth*, economist Juliet Schor states:

"Since 1990, inflation-adjusted per-person expenditures have risen 300 percent for furniture and household goods, 80 percent for apparel, and 15–20 percent for vehicles, housing and food. Overall, average real per-person spending increased 42 percent" (2010: 6). Not surprisingly, we have needed to increase our home size to accommodate all of this stuff. Schor cites the U.S. Census Bureau when she relates that in 1980 the average single-family home was just over 1,700 square feet. By the start of the twenty-first century the average single-family home had increased by 45 percent, to just over 2,500 square feet (45).

Even an ever-expanding home size hasn't, however, been able to handle all of our acquisitions. In order to increase our consumption at such astonishing rates, we have also had to learn to get rid of lots of stuff. Schor shows that increasing rates of U.S. clothing imports are matched by increasing exports of secondhand clothing. She estimates that in 1991, 316 million pounds of used clothing were exported globally from the United States. By 2004, exports were 1.1 billion — almost four times as much (39).

It is, however, the electronics industry that I think best exemplifies how incredibly adept we have become at coveting shiny new objects (and trashing perfectly good earlier versions to make way for the newest model). The U.S. Environmental Protection Agency (EPA) estimates that in 2010, Americans purchased just over 23.5 million desktop computers and 40.4 million laptop/portable computers. At the other end of the computer lifecycle, the EPA estimates that Americans discarded 51.9 million desktop and laptop computers in 2010 — or more than 142,000 computers discarded *every day*! It is no surprise that no-longer-desired computers and other electronics — also known as "e-waste" — represent the fastest growing municipal waste stream in the United States (EPA 2008).

Lest you are tempted to think that the good old television is somehow exempt from this phenomenon of electronics consumption and purging, consider these statistics. The EPA estimates that approximately 34 million televisions were purchased and 28.5 million television sets were disposed of in the United States in 2010 (that's about 78,000 sets discarded every day). Want to hazard a guess as to when there's an upward blip in American's television purchasing every year? The Super Bowl. Yes, according to the National Retail Association, in 2008, Americans bought just shy of four million "conventional" television sets in honour of the Super Bowl (up from the 2.5 million Super Bowl television sets purchased in 2007, which was up from the 1.7 million Super Bowl television sets purchased in 2006). According to the Consumer Electronics Association, we need to add to the above numbers the high definition (HD) TVs that were purchased for the Super Bowl — 2.6 million in 2009 and 3.3 million in 2010.

It's Not Just the Ads

We have learned our materialism lessons well, and why wouldn't we? As Postman observed: "An American who has reached the age of forty will have seen well over one million television commercials in his or her lifetime. We may safely assume, therefore, that the television commercial has profoundly influenced [North]American habits of thought" (1986: 126). While Postman's observation was made over twenty-five years ago, it continues to be prescient and astute. But here's the thing about linking advertising and materialism: it's easy to stop there. With the reams of research, not to mention common sense, it's easy to link advertising's constant bombardment with the exemplary materialists we have so quickly become. If we leave the analysis focused on advertising, however, we would be making a huge mistake. No doubt the "classic advertising" to which Postman alludes has played a role in shifting our relationship with the material world. Advertising historians will remind us just how hard the ad men (and they were almost exclusively men at first) worked to shift that relationship. But as Raymond Williams states in *Culture and Materialism*, advertising was not content to be isolated.

> An out-dated and inefficient kind of information about goods and services has been surpassed by the competitive needs of the corporations, and these increasingly demand not a sector but a world, not a reservation but a whole society, not a break or a column but whole newspapers and broadcasting services in which to operate. (2005: 195)

Advertisers have historically relied on the explicit 30- or 60-second "commercial" to help teach people how to be good materialists, but it was only a matter of time before they embraced the reality that *everything* on television could sell.

Some point to the 1982 movie *E.T.: The Extra Terrestrial* as the moment when "product placement" was born. In the movie, ten-year-old Elliott scatters the small brightly coloured peanut-chocolate candies around the forest in the hope that they will attract an alien he has seen. The ploy works (E.T. actually collects up the candies and returns them to Elliott), and Elliott is able to lure E.T. into his home using the candy. The two become friends and eventually Elliott helps E.T. return home. The use of the Reese's Pieces candy was a result of a deal with the candy manufacturer Hershey's Foods. In return for this in-film promotion, Hershey Foods did a million dollars of advertising for the movie. It would seem that Reese's Pieces was the big winner in the deal, however, because, according to Barbara Mikkelson, writing for snopes.com, "Within two weeks of the movie's premiere, Reese's Pieces sales went through the roof. (Disagreement exists as to how far through the roof they went: sales were variously described as having tripled, experienced an 85% jump, or increased by 65%)."

Whether the sharing of Reese's Pieces with E.T. is the most accurate place to mark the start of product placement is up for debate, but certainly ever since that silver screen moment, product placement — and the more narratively embedded "product integration" (i.e., rather than just showing the product, the characters actually *talk about the product*) — have been alive and well. For example, Nielsen has been tracking product placement by counting the number of times a product appears in a show's segment. From January 1, 2011, to November 30, 2011, the top three product placement shows on network television (ABC, CBS, CW, FOX and NBC) were *American Idol*, with 577 product placements, *The Biggest Loser*, with 533, and *The Celebrity Apprentice*, with 391 (not surprisingly, for anyone who has watched reality television, nine of the top ten product placement spots were held by reality television shows).

In the late 1990s, researchers Rosellina Ferraro and Rosemary Avery carried out a content analysis of 112 hours of prime-time television programming (7–11 p.m.) on the four major U.S. networks (ABC, CBS, NBC and FOX) in order to look at the types of products/brands that are shown and/or talked about during "non-advertising" programming (situational comedies, made-for-television movies, dramas/series, comedy skit/variety, cartoons and movie reruns). The researchers found that the average verbal and visual references to specific products and brands ranged from a low of 7.5 per 30-minute segment for CBS to a high of 18.4 per 30-minute segment for NBC. Sports programming had the highest number of product and brand references (58.2 per 30-minute period), followed, interestingly perhaps, by long format news programming/feature magazines (40.6 references per 30-minute programming) and news (36 references per 30-minutes of programming) (Ferraro and Avery 2000).

However, while findings such as these are striking and remind us that we need not wait for the advertisements to have a product promoted to us, I propose that materialism's more powerful and profound lessons are woven more subtly into the fabric of television's stories. Think about the fact that television does not have 30-second advertising spots in the name of promoting racism, sexism, ageism and violence, yet cultivation theory research clearly shows that television affects our understanding of these concepts. So while television does offer 30-second advertising spots in the name of explicitly promoting materialism, cultivation research highlights that television's role as storyteller promotes materialism far beyond advertisements or product placement. Cultivation researcher L.J. Shrum points out: "When it comes to the formation and maintenance of social beliefs, television *programming* should not be overlooked in an attempt to understand consumer socialization" (O'Guinn and Shrum 1997: 291, italics added).

Television and Materialism

There is no better teacher of materialism than television. Indeed, the explicit promotion of materialism (through advertising, product placement) obfuscates television's real power in this regard. Why is television such a good teacher? Because, as I detailed in the previous chapter, we have welcomed television into our homes such that the televisions outnumber the people found there. It is not uncommon to find a television in our bedroom and our children's bedrooms, and its flickering light has been known to lull us, and our children, to sleep. Only sleep and work/school compete with television viewing as the activity on which we spend the most time each day. In other words, we *love* our television and the stories it tells us. Advertising, product placement and product integration are television's explicit materialism moments, but television's true power is contained in the reality that television is, as Shanahan and Morgan propose, a "cultural river, in which everyone to some degree is carried along" (1999: 12).

Cultivation research highlights that television has taught us all kinds of things about the world, including that it is a mean and scary place, that men outnumber women, that women perform gender-stereotyped roles, that people of colour commit most of the crimes while Caucasians solve most of the crimes, that women's bodies should conform to very limited physical parameters, that the elderly are invisible and that we should believe that what we own is who we are (Shanahan and Morgan 1999).

Interestingly, while there is research on all of these topics, especially the relationship between television and fear about the world as a violent place, when it comes to the relationship between television and materialism, the research is limited. For example, if we do a simple analysis of the Communication and Mass Media Complete database (a collection of about 820 communication and mass media academic publications), the dearth of studies that have explored the links between television and materialism is clear. A search on articles related to "television" and "violence" published between January 1958 and October 2012 yields just over 1,200 results. By contrast, a search on articles related to "television" and "materialism" yields thirty-eight results. As Mark Harmon points out, "few researchers... have taken the path that the most consistent and significant message of commercial television is commercialism and that the place to look for long-term, subtle, and pervasive effects is in materialism among heavy [television] viewers" (2001: 406). L.J. Shrum has similarly commented that materialism is "a personal value construct that has received little attention in cultivation research" (Shrum et al. 2011: 35).

While studies on the relationship between television and materialism have been few and far between, there have been some. One of the earliest examples of research exploring the links between television and materialism is Gilbert Churchill and George Moschis's 1979 article "Television and Interpersonal Influences on Adolescent Learning." The researchers found that there was a relationship between adolescents' higher levels of television viewing and more positive attitudes about materialism. Richard Easterlin and Eileen Crimmins (1991) also used a sample of teenagers/young adults to explore the relationship between overall television use and materialism. While their secondary data analysis (i.e., analyzing data others had collected) yielded inconclusive results, a study by Chau-Kiu Cheung and Chi-Fai Chan five years later, using a sample of adolescents from Hong Kong, corroborated Churchill and Moschis's findings that those who watched television tended to be more materialistic.

In 1998, M. Joseph Sirgy and colleagues found a relationship between television viewing and life satisfaction, including materialism, for five countries (China, Turkey, Australia, Canada and the United States). In his article "Affluenza: Television Use and Cultivation of Materialism" (2001), Mark Harmon explored the possible relationship between television viewing levels and materialism. Using a large-scale American survey, he found that those respondents who regarded "nice things as important" and " a high income as important" were heavier viewers of television.

Other television and materialism researchers have made use of specific types of television viewing rather than overall viewing. For example, Bo Reimer and Karl Rosengren (1990) found that Swedes who watched fiction/ entertainment programming were more likely to have materialistic attitudes, whereas those who watched non-fiction (news, "high culture") were more likely to have non-materialist attitudes. In their 1994 article "Commercials in the Classroom: The Impact of Channel One Advertising," Jeffrey Brand and Bradley Greenberg explored the relationship between materialism and Channel One, a company that shows students an advertisement-laden daily ten-minute television news broadcast at school in return for which the school receives monetary and material rewards. The researchers found a relationship between school children's exposure to Channel One and higher levels of materialism.

In the 1990s, L.J. Shrum and his colleagues compared heavy and light television viewers who were consuming television content that was particularly materialistic. Based on content analyses, Thomas O'Guinn and Shrum decided to use daytime soap operas as a materialistic genre. The researchers created a survey that asked participants to estimate the percent of Americans who owned expensive items (jewelry, cars, fur coats) or participated in affluent leisure activities such as going to the spa, playing golf and drinking wine.

The survey was then given to two groups of women. One group had indicated that they were heavy viewers of soap operas. The other group had indicated that they did not watch soap operas at all. O'Guinn and Shrum found that heavy viewers of soap operas were more likely to overestimate rates of societal wealth and consumption, leading the researchers to conclude that "television *programming* is a significant, yet overlooked source of consumption-related social perceptions" (1997: 291, italics added).

Shrum and his colleagues followed up on their soap opera research in 2005 with an article entitled "Television's Cultivation of Material Values." In this study, participants were asked for their level of agreement with various comments about materialism based on a fifteen-statement scale developed by Richins and Dawson (similar to their eighteen-statement materialism scale, discussed earlier in the chapter). The participants were also asked about their television viewing based on a six-statement scale developed by Shrum and his colleagues (see below). The researchers discovered that there was a relationship between heavier levels of viewing television and higher scores on the materialism scale.

In cultivation research that I conducted, I built on these materialism studies of Shrum and others. Part of my data collection involved sending a survey to a thousand randomly chosen Americans. One of the scales I used was Richins and Dawson's eighteen-statement materialism scale. Participants responded to the statements with a 1 to 7, where 1 = strongly disagree and 7 = strongly agree. I also asked participants about their television viewing habits. I did this in two ways. First, I asked about average weekday viewing and average weekend viewing (these can be quite different for some people). Second, I made use of a television viewing Likert scale that Shrum and his colleagues developed (see Shrum et al. 2005; 2011). This scale's six statements are as follows (note that this scale has some "reverse-coded items," but the scale would be coded such that higher scores would equal heavier television viewing).

1. I watch less television than most people I know.
2. I often watch television on weekends.
3. I spend time watching television almost every day.
4. One of the first things I do in the evening is turn on the television.
5. I hardly ever watch television.
6. I have to admit, I watch a lot of television.

Once I had responses to the materialism and television viewing scales, I compared the level of television viewing with the materialism score (statistically controlling for such things as income, gender, age and whether someone lived in a rural, urban or suburban area). Similar to the research highlighted above, I found that participants who watched more television, on average, also tended to score higher on the materialism scale.

In the most recent cultivation of materialism research, Shrum and his colleagues undertook the task of not only illustrating the relationship between television and materialism but also of refining our understanding of that relationship. The researchers found that television viewing is related to materialism, and that "the relation is robust" (2011: 47), and were able to show that television caused the changes in materialism and explain why.

Which Comes First?

Almost all of the research I highlight points to a relationship between higher levels of television viewing and higher levels of materialism, but does television viewing *cause* higher levels of materialism? Or, perhaps, do those who are more materialistic tend to be drawn to television's content?

One way Shrum and his colleagues explored whether materialistic people are somehow more drawn to television is by experimentally manipulating the degree of materialism in the television content. In a study published in 2005, the researchers showed participants two movie clips — one from *Gorillas in the Mist* (pretested with research participants who rated the clip "non-materialistic") and one from *Wall Street* (pretested with research participants who rated the clip "materialistic"). They found that participants (who ranged in their levels of materialism) enjoyed the television content equally regardless of whether the content was high in materialism or low in materialism. Or, in other words, people watch television for its entertainment value, and regardless of how materialistic a person is, the materialistic content is, according to the researchers, "a by-product" of, and *not the reason for*, deeming television content entertaining.

In a subsequent study, using 142 university students, Shrum and his colleagues (2011) made use of the same stimulus viewing content, clips from *Gorillas in the Mist* and *Wall Street* shown on television. The findings demonstrated those who had been randomly assigned to watch *Wall Street* had higher levels of materialism immediately after viewing than did those who had been randomly assigned to watch *Gorillas in the Mist*. Taken together, Shrum's research indicates that the direction for the relationship between television and materialism is *from television to materialism*, or that television is causing greater materialism *only in heavy viewers* — and not that materialistic people are drawn to watching greater quantities of television. This of course leads us to the next question: Why?!

Why Does Television Make Us Materialistic?

Poet Muriel Rukeyser proposed that, "The universe is made of stories, not of atoms." Our stories create our reality. Television is our storyteller, and no story television tells is as omnipresent, or as powerful, as how we should understand our relationship with the material world. Cultivation research demonstrates that not only does television tell us about materialism, heavy viewers of television are also more likely to internalize those stories and become even more materialistic.

Studies of television's content illustrate that we can count the blatant "incidents" of materialism on television. We can measure, as Ferraro and Avery did, the frequency of in-program product references. We can look, as Shrum and his colleagues did, at the types of affluence that appear in a particular type of programming (e.g., soap operas). But the real power is in television's overarching message, or "meta-message." The message in television as storyteller is much more subtle and implicit. To help highlight this subtle implicitness, let me introduce you to the concept of symbolic annihilation.

Symbolic annihilation is a concept that cultivation researchers developed to explain that what we learn from television is not only based on what television shows us and tells us but also on what television *doesn't show us* and *doesn't tell us*. For example, the elderly are a demographic that television rarely shows us, and research indicates that heavier television viewers are more likely to have misconceptions about the elderly (e.g., how prevalent they are in society) than are lighter television viewers (Shanahan and McComas 1999). Similarly, while television explicitly tells us things about how we should relate to the material world — what things we should covet, how our homes should look, how we should dress, how we should relate to (and long for) money — television also tells us about the material world through what it *doesn't show us*. When television doesn't tell us stories that challenge materialism, then television *symbolically annihilates* those challenges.

In its more subtle form, television symbolically annihilates by simply not telling certain kinds of stories. In general, for example, television does not tell us stories that revolve around the poor and working poor — people at the centre of television's stories are most often affluent. Similarly, there are few unaesthetically pleasing characters — people on television most often aspire and conform to mainstream societal definitions of physical and fashion attractiveness (unless they are evil; evil people can be unfashionable and unattractive). Television does not show us people who share their belongings or find other ways to actively resist consumption. Similarly, consider what people compete for in television contests; it is not usually the opportunity to make a charitable donation or better someone else's life (an exception was Oprah Winfrey's *The Big Give*, in which contestants gave money away and were in competition to win a million dollars — half of which they gave to charity). Game shows and the more recent phenomenon of reality television almost always give away cash, cars and makeovers to individuals willing to debase themselves in a variety of ways. These examples are the more subtle ways in which television symbolically annihilates: television tells stories that include some ingredients (the desire to have more money and stuff) and not others (being economically disadvantaged, unattractive, satisfied with "enough").

But there are also explicit examples of how television symbolically annihilates. In the late 1980s, Canadian lumber giant MacMillan Bloedel ran a Canadian Broadcasting Corporation (CBC) television advertising campaign called *Forests Forever*, aimed at reassuring viewers that Canadian forests were being carefully managed (Wilson 1998). Two documentary filmmakers, Kalle Lasn and Bill Schmaltz, decided they didn't agree with the "we're managing the forests just fine" message and put together their own television advertisement questioning the sustainability of Canada's forestry practices. These practices include clearcuting huge tracts of original forest and replacing them with what Lasn and Schmaltz, and other environmentalists, call "tree farms" (Adbusters 1993).

When Lasn and Schmaltz went to purchase airtime for their television advertisement, however, the CBC refused to run the "un-commercial" (as the activists called their ads). Why did the CBC refuse? Because they deemed the un-commercial "too controversial." A movement was born from the ensuing questions: "Controversial for whom?" and "Who is allowed to buy television advertising time on the public airwaves — and who isn't?" From the media attention that the un-commercial and general debate garnered, several activist endeavours were launched, including *Adbusters* magazine, the term "culture jamming," additional "un-commercials" and International Buy Nothing Day (Adbusters – Cuture Jammer's Manifesto).

Explicit symbolic annihilation therefore occurs when advertising and programming that celebrate materialism are allowed on television, but advertising and programming that question and oppose materialism are not welcome. In a taped phone exchange (available on the *Adbusters* website), a television representative says: "Why would we air these commercials? It would be bad for business." It is hard to deny the logic. Why would a company that exists to sell the culture of materialism allow messages and stories that question that culture?

Symbolic annihilation is one way of making sense of how watching television makes us more materialistic, but it explains television's influence in connection with what is *not* on television. The research into the cognition of the cultivation effect helps us make sense of how the stories that *are* told on television become "our stories."

As part of the exploration of cultivation theory (in the previous chapter), I looked at the question of *why* there is a cultivation effect. I explored why, when we watch a lot of television, our understanding of what the world is like increasingly mirrors the world shown on television. I talked about the fact that while many cultivation researchers have not been particularly interested in the "Why?" part of cultivation, one researcher in particular, L.J. Shrum, has investigated this question.

In the Television chapter, I introduced the concept of heuristics, or the mental short-cuts that we make use of as we cognitively process what is going on around us. The idea is that because we can't exhaustively think through all of the stimuli we experience, we rely on mental short-cuts. For example, let's say a friend is leaving your house one night, and as she leaves she asks you if it's safe to walk alone. Your friend wants you to assess what the chances are that something bad will happen to her if she walks alone. Shrum proposes that when we are asked to make these kinds of "what are the chances"-type judgments, we draw on information about the topic that is most readily available to us. In this case, you will quickly scan what comes to mind when you think about concepts such as "crime," "safety at night," "my neighborhood," etc. If you've ever had a bad experience walking alone at night or know someone who has, that would certainly come to mind. What else will probably be part of your calculation, especially if you watch a lot of television? You'll consider bad nighttime experiences you've "experienced" on television (and chances are that the "television experiences" will be added into your equation of "safe or not?" without you being aware that you're doing it). In other words, as we develop these mental short-cuts to assess things like "what is safe?" we grab onto what is most cognitively available — especially if we are trying to process an answer quickly. And if we watch a lot of television, what we have experienced via television is easily and quickly available.

As I pointed out in the Television chapter, one of the ways in which Shrum's research highlighted this "availability heuristic" is by showing that when people are encouraged to be thoughtful in their answers, television's information is more likely to be discounted. In fact, if a survey asks about television viewing habits first and then asks about judgments, estimates, etc. second, the effects of television viewing are removed altogether. This is because television's role in our thinking has been highlighted, and we are able to discount that information and focus on what we know from our lived experience. Shrum has similarly shown this "thoughtfulness effect" by having research participants respond to questions via a snail-mail survey or via a phone-administered survey. As you may recall, the phone survey participants showed a higher cultivation effect because they had less of an opportunity to dismiss their "television answer" than did the snail-mail survey participants (Shrum 2007).

How does this "thoughtfulness effect" relate to television and materialism? One way it relates is "resonance." As I mentioned in the Television chapter, resonance happens when television's messages are supported or corroborated by our lived experience, resulting in the cultivation effect being that much stronger. This means that when television's omnipresent materialism is met by other messages of materialism (on billboards, in our

Facebook activities, in our friend's wishes for a new smart phone — arguably everywhere in our lives), then there is near-constant reinforcement of television's materialistic stories.

There is another way that Shrum's concept of heuristics relates to television's relationship with materialism. Cultivation theorists propose that there are two categories of television effects — first-order effects and second-order effects. First-order effects refer to the fact that heavier viewers of television are more likely to draw upon television to answer questions that involve judgments about such things as the likelihood of something happening (e.g., a friend asking about the safety of walking home alone). Second-order effects refer to the fact that heavier viewers of television are more likely to draw upon television to answer questions about our values, morals and beliefs.

Why do these first-order and second order effects matter? The answer has to do with *when television has its influence*. Shrum proposes that because we don't tend to carry estimates and frequencies around with us, we develop them in real time (i.e., when our friend asks about the safety of walking alone at night, we probably don't have an answer immediately ready but piece together an answer based on the available heuristics). As such, television's first-order effects happen as our answers are being crafted (i.e., after the television viewing has taken place). Unlike estimates and frequencies, however, our values, morals and beliefs are with us at all times. When we are asked whether we think killing another person is wrong, or whether the more money one has the more successful one is, we have a gut-level feeling ready to help us answer the question at any time. Shrum therefore proposes that television's second-order effects happen *while* we watch television.

Why does it matter *when* television's stories affect us? One area that Shrum has explored involves the implications for how much we like to think. We know that the opportunity for thoughtfulness can suppress the cultivation effect. Consider what happens when someone is particularly thoughtful. Perhaps that person will be less "susceptible" to the cultivation effect. Shrum's findings are surprising.

In the article "Television's Cultivation of Material Values," Shrum and colleagues made use of the Need for Cognition (NFC) scale to measure how thoughtful a person is and how much they like to think. The scale asks research participants about their level of agreement (strongly agree to strongly disagree) with statements such as "I prefer my life to be filled with puzzles that I must solve," "Thinking is not my idea of fun," and "The notion of thinking abstractly is appealing to me." What the researchers found was that while there was a relationship between heavier television viewers and higher scores on the materialism scale, heavier television viewers who *also* had higher scores on the NFC scale tended to have the *highest* materialism scores. Thus the researchers concluded that "the effect of television viewing on material values was greater for those who reported that they tend to pay more attention while viewing and who tend to elaborate more [or think more about the content] while viewing" (Shrum et al. 2005: 477).

These findings are perhaps unexpected and raise an interesting conundrum for anyone wishing to counter television's effects on materialism. The conventional wisdom might be that, well, wisdom can help. Perhaps if we're thoughtful enough, we can be aware of how we're being materialistically appealed to by television and think our way past those appeals. But the research indicates that the smarter we are — the more we like to engage ideas and think about the world in which we live — the more likely we are to also engage television's ideas and think about television's world as we watch. After all, people who watch a lot of television do so because, presumably, they like it. Even thoughtful people who watch television for hours each day don't want to spend their time deconstructing everything that comes across the screen. No, if the TV is on for hours a day, it's because the content is enjoyable, and when a thoughtful person is watching enjoyable content, the content is being even more intently and actively engaged.

That said, as I pointed out in the Television chapter, the cultivation research on television and materialism, like all cultivation research, does not show huge statistical power. Television, therefore, cannot explain everything about how we've come to understand materialism. What the research does consistently indicate, however, is that television adds to our understanding of the material world along with the other major socializing forces — family, friends, school, work, faith groups, sports teams, etc. Keep in mind, however, that when we learn about the world from family, friends, workmates and members of our faith groups and sports teams, we're learning from people who have also — almost without exception — been steeped in television's stories. Television serves as a kind of bio-accumulator of our materialistic understanding of the world.

How to Make Sense of Materialism

Perhaps at this point you're thinking "Great!" and "Good for television!" It's a pretty good deal, really. Television viewers get to sit and be entertained (quite well entertained it would seem — otherwise why consume its content for hours each day?) and the marketplace can go about its business of selling. As my students are always quick to point out, we can understand television's materialistic stories (explicit and implicit ones alike) as having great utilitarian value: television's stories provide us with the specifics of product information and more generally help promote the desirability of stuff. Who cares if kids watch television and want to play with the toy they see? Perhaps they'll even convince a parent or grandparent to purchase the toy and lubricate the economy a little in the process. So what if heavy viewers of television are more likely to have their sense of self intimately connected with the things they own? Buying stuff can make us feel good about ourselves (you've heard of "retail therapy"); our stuff makes us happy, right? After all, we live in a society that is intimately connected to the marketplace, and if television is an effective conveyor of what's in that marketplace and encourages us to purchase what's available, then good for television, right?

One way that people make sense of the television-materialism connection is by pointing out that we (via entertainment and information) and the market economy (via us being encouraged to buy stuff) benefit from materialism's synergistic relationship with television. Another common counterpoint to a critique of television's role in encouraging materialism runs along the lines of, "Well, we don't want to go back to living in caves and eating berries, do we?" In fact, one could argue that if subsistence cave dwelling and berry eating is at one end of the "ways we can live" continuum, then the access to the world of plenty that twenty-first-century materialism offers is at the other end. Modern-day materialism is about access to food, shelter and possessions so far beyond subsistence that we not only purchase these essentials because we need these things to survive, but we also pick and choose our food, shelter and possessions to *impress others with what they say about us*. How incredibly evolved is that?!

Perhaps we should understand materialism, and thus television's role in fostering materialism, as a kind of pinnacle. The Industrial Revolution's urban centres, to which we were drawn to work in the factories, and the subsequent necessity of teaching us to be consumers, have been ultimate victories of human ingenuity and progress, no? Have we not been freed from the shackles of subsistence and the fear of want that plagued generations of our ancestors? Should we not celebrate our ability to dream but to also own that of which we dream? Should we not sing the praises of the economic system that brought us this world of plenty?

The answers to these questions are complicated. Yes, in the more affluent parts of the world, it is easy to see freedom from want and unprecedented access to material goods. But at what cost? Richard Ryan notes in his foreword to Tim Kasser's book *The High Price of Materialism*:

> At this point in human history we have enough material resources to feed, clothe, shelter, and educate every living individual on Earth. Not only that: we have at the same time the global capacity to enhance health care, fight major diseases, and considerably clean up the environment. That such resources exist is not merely a utopian fantasy, it is a reality about which there is little serious debate. Nonetheless, a quick look around most any part of this... globe tells us just how far we are from achieving any of these goals. (2002: ix)

But let's not enter into a discussion of the inequitable distribution of material wealth yet. Let's keep our focus on those seemingly lucky ones in the more affluent parts of the world. At least for them materialism has been a shining success, hasn't it?

In *The High Price of Materialism,* Kasser offers the details of his ongoing materialism research as well as elements of over 200 other materialism studies. The essence of this research can be divided into two basic steps. First, a materialism researcher needs a way of exploring and often (although not always) quantifying the research participants' materialism. I highlight some of the varied ways in which this has been done in what follows, but the essence is that research participants are asked to respond to statements about what they value, strive for and aspire to be in their lives. Second, a materialism researcher needs a way of exploring and often (although not always) quantifying the research participants' happiness/well-being. I also highlight some of the ways in which that has been done, but basically research participants are asked to either offer their own sense of "how much" happiness/well-being they have and/or they are interviewed by a mental health professional, who provides an "objective assessment" of the participant's happiness/well-being.

Kasser highlights that the early materialism researcher Russell Belk (writing in the 1980s) tapped into materialism by asking participants questions related to three components of materialism: possessiveness, non-generosity and envy. Belk explored participants' well-being by simply asking two questions: "How happy are you?" and "How satisfied are you with your life?" What he found was that those who scored higher in the three components of materialism scored lower in well-being.

In the early 1990s, Marsha Richins and Scott Dawson's research drew upon and tweaked Belk's approach to studying materialism and well-being. As with Belk, Richins and Dawson's measure of materialism focused on people's interest in having possessions and money, but the researchers expanded the measure to include people's interest in impressing others with what they own. Richins and Dawson's well-being measures were also similar to Belk's, asking people about their general happiness, but they expanded this measure to include the specifics of happiness regarding the participants's home and work life. (An example of their materialism scale can be found above in "The Meaning of Materialism" section of this chapter.) Their findings? Similar to Belk's research, Richins and Dawson found that the higher a person's interest in material possessions/money and impressing others, the lower their well-being.

Tim Kasser's initial research, also in the early 1990s, which he conducted with his colleague Richard Ryan, looked at the relationship between people's financial aspirations and various indicators of well-being. Their initial measure of "financial aspirations" included asking research participants about the desirability of a future that included elements such as, "You will buy things just because you want them" and "You will have a job with high social status." Their measures of well-being included survey responses and assessments of mental health by a psychologist. The two sides of the equation (materialism and well-being) were then compared. What Kasser and Ryan found in these early studies, as Kasser summarizes in *The High Price of Materialism*, was that "when young adults report that financial success is relatively central to their aspirations, low well-being, high distress, and difficulty adjusting to life are also evident" (2002: 9).

Over the years there have been many other ways that researchers have tried to tap into the correlates, what "goes with," materialism. Patricia and Jacob Cohen, for example, had 700 participants who were aged twelve to twenty respond to questions about materialism and undergo an interview with a psychologist. The researchers found that those participants who scored higher on materialism (including the goal to one day "be rich") were also more likely to be diagnosed with various mental health disorders, including alcohol abuse, depression and narcissism (cited in Kasser 2002: 14–17).

Kasser and Ken Sheldon explored materialism by asking research participants to list their life goals and then "categorize" those goals according to six futures: three materialistic futures — financial success, fame/popularity, physical attractiveness; and three non-materialistic futures — self-acceptance/ personal growth, intimacy/friendship, societal contribution. Regardless of age or gender, participants' "materialistic values appear not to bring happiness and wellbeing, but instead more anxiety, little vitality, few pleasant emotions, and low life satisfaction" (Kasser and Sheldon 2002: 22).

With this wealth of materialism research in mind, therefore, Kasser offers that "the value of materialism yields clear and consistent findings. People who are highly focused on materialistic values have lower personal well-being and psychological health than those who believe that materialistic pursuits are relatively unimportant" (2002: 22). Yet, as with other bad habits in which we might participate despite knowing they are bad for us — smoking, over-eating, not exercising — many of us continue to believe that our stuff makes us happy and that we are what we own. We continue to want more things, newer things, "better" things, in spite of our ability to recognize, at some fundamental level, that we own everything we need... that more stuff will put us further in debt... that debt plays a role in keeping us from doing things that really do make us happy, like spending more time with family and friends.

So *why* do we do it? How did we get here? As I've highlighted above, I think the answer lies in the stories we're told about ourselves and about our world, most importantly the stories we're told by television. But there's still a question about why we're *susceptible* to those stories. Why, when we have the material possessions that we need and that make us comfortable, are we still so open to the notion that we should want more? Is there a logic to materialism?

Insecurity

It makes intuitive sense that our desire to acquire would be related to what we lack — that we'll want more strongly that which we do not have. Food is an easy example. If you have ever been hungry, even temporarily, and even due to your own volition (for example, while on a camping trip or on a diet), you know how easy it is to become obsessed with food. For those who are regularly hungry due to factors outside of their control, thinking about and pursuing food is an obsession. Psychologist Abraham Maslow noted: "For the man who is extremely and dangerously hungry, no other interests exist but food. He dreams food, he remembers food, he thinks about food, he emotes only about food, he perceives only food and he wants only food" (1943: 174). But in the twenty-first century we also know that easy access to plentiful sustenance does not necessarily mean that our obsession with food ends. Far from it.

While the United Nations' World Food Programme estimates that globally 925 million people are undernourished, at the other end of the continuum the World Health Organization (2012) estimates that globally as of 2008 there were almost a billion and a half overweight adults (twenty years and above) and more than 500 million obese adults. So, while it makes perfect sense that those who are hungry become obsessed with the acquisition of food, the picture is complicated by those who have essentially unlimited access to food and continue to be, if not obsessed, then at least occupied with and interested in food to the point of being overweight or obese.

Don't get me wrong, I understand that we have a hugely complex relationship with food, and I also understand that food and material possessions are not the same, but there are fundamental similarities. As with food, our survival depends on certain essential material resources — such as clothing and shelter — and in their absence there can be desperation. So, while the absence of material fundamentals may not be directly related to physiological need in the way that food is, we can understand the central importance of material possessions in our lives. Indeed, Maslow (1943) is best known for his proposition that we have a hierarchy of needs such that our fundamental physiological (e.g., breathing, food, water) and safety (e.g., physical health, sense of security, a place to call home) needs are the foundations on which everything else in our lives is built.

Maslow came to his conclusions about our hierarchy of needs after he studied those considered to be the "most successful" in society (public figures as well as average citizens) and proposed that we can only move to the higher levels of "self-actualization" (solving problems, being creative, pondering morality, etc.) once all of our fundamental needs have been met. Material possessions, it can therefore be argued, facilitate one's ability to move from the base of Maslow's hierarchy. If we are without food, if we don't have somewhere to sleep or go to the bathroom, we are stuck at the bottom of the hierarchy only able to gaze at, and covet, what those "above" have — food, shelter and beyond this a sense of belonging, being loved, self-esteem and self-actualization.

Here's the dilemma that Maslow did not anticipate. We now know that even when our physiological needs have been met and our lives contain great material abundance, we often eagerly, and endlessly, covet and seek to acquire more (as in the food analogy). In fact, one could say that in the more affluent parts of the world we have increasingly gotten stuck in the bottom two layers of Maslow's hierarchy — obsessed with what he called our physiological and safety needs — and are often unable to move up into the layers of belonging, love, self-esteem and self-actualization. Why? Economist Juliet Schor points out that "human nature" has been proposed as an explanation for what she calls our "maximally acquisitive behavior" but that "the evolutionary psychology literature supporting this position falls short of being convincing" (2010: 170).

No, the answer for why we have gotten stuck endlessly acquiring more stuff has to do less with some kind of deep-seated human nature and more to do with insecurity. Insecurity has always been at the heart of materialism. As was highlighted earlier, advertisers learned early that the key to selling was convincing potential buyers that they should be insecure and understand themselves as incomplete without the advertised product. "It was recognized," Ewen writes, "that in order to get people to consume and, more importantly, to keep them consuming, it was more efficient to endow them with a critical self-consciousness in tune with the 'solutions' of the marketplace than to fragmentarily argue for products on their own merit" (1976: 38). Tim Kasser has undertaken some fascinating research that illustrates the link between insecurity and materialism.

First, Kasser and his research colleagues administered a materialism scale, called the Aspiration Index (AI), that measured participants' financial aspirations. The AI contained the following six future life goals, which participants would rate from "extremely important" to "not at all important":

1. You will be physically healthy.
2. Your name will be known by many people.
3. You will have people comment often about how attractive you look.
4. You will have a lot of expensive possessions.

5. You will be famous.
6. You will donate time or money to charity.

Kasser and his colleagues then selected their research participants for the materialism study from the top 10 percent and bottom 10 percent of the respondents to the AI scale.

In order to tap into subconscious cognitive processes, Kasser and his colleagues then asked these high and low scoring AI individuals to talk about their dreams. The researchers found that those who scored higher in materialism were also more likely to report dreams that involved death and falling (two fears that are, for most of us, fundamentally about insecurity).

As with all research that shows a correlation (in this case between fearful dream content and materialism), it was not clear what caused what. In other words, are fearful people more likely to be materialistic, or are materialistic people more likely to be fearful? In a fascinating follow-up study, Kasser and his colleagues sought to experimentally test this question of *what was causing what*. The experiment involved randomly placing university students in one of two conditions.

In condition one, students were asked to write two essays: one essay that "describes the feelings that the thought of your own death arouses in you" and another essay that "describes what you think will happen to you physically as you die and once you are dead." In condition two, students were also asked to write two essays, but instead of death, the topic was music. The students were asked to write one essay that "describes the feelings that the thought of music arouses in you" and another essay that "describes what happens to you physically as you anticipate listening to music and as you listen to music." Keep in mind that by randomly placing participants in the two conditions, there were no significant differences between the members of the two groups prior to writing the essays.

After writing the essays (either about music or death), the participants were asked to play a game in which they were the managers of a hundred-acre forest. As forest managers, the participants had to decide on three aspects of forest management:

1. How much *more money* they wanted to make from their forest than other forest companies (testing how *greedy* the participants were).
2. How much they worried that these other forest companies would chop down the forest *before they did* (testing how *fearful* they were).
3. How much of the hundred acres of forest they would chop down (testing how *consumption-driven* they were).

On average, those who wrote about death wanted to chop down sixty-two acres, while those who had written about music wanted to chop forty-nine acres (i.e., writing about death had made the participants more consumption-driven). The participants who wrote about death were also greedier — expressing a greater interest in making more money than the other companies. Since the participants were randomly assigned to either condition (i.e., neither group was more materialistic before writing the essays), then we can assume that writing about death invoked a sense of insecurity and this insecurity made the participants more materialistic — and this materialism explains the differences in their forest management practices.

Now you might be thinking, "So what?" After all, it's not like anyone is encouraging us to sit around writing essays about death and thereby instilling feelings of insecurity in us. But there is a way that insecurity is more generally, and less bleakly, fostered: through comparison. In other words, we may feel pretty good about ourselves and our life situation until someone comes along and says, "Look, here's a situation that's *better*." Suddenly we're insecure. Suddenly we're wondering, "Might my situation, which seemed fine until now, be *better* than it is?" When the difference between our situation and the "other" situation appears to be a difference of material possessions, the coveting begins.

Here's a fascinating example of this phenomenon. This case involves the South Pacific island nation of Vanuatu where, as I mentioned above, I began writing this book. Vanuatu is comprised of eighty-two islands and is blessed with both a moderate climate and abundant natural resources. Historically, therefore, the land has provided well for the indigenous ni-Vanuatu people. European explorers first arrived in the 1600s, but the majority of ni-Vanuatu people had very little contact with foreigners, or their fancy imported possessions, in the centuries that followed. Then in 1942 the Second World War brought unprecedented numbers of foreigners and huge quantities of their stuff to Vanuatu's shores.

Several of Vanuatu's islands became strategic posts for the American war effort. The ni-Vanuatu, who had previously known only subsistence hunting and gathering, traditional huts and firelight, were introduced to "radios, TVs, trucks, boats, watches, iceboxes, medicine, Coca-Cola and many other wonderful things," and as Paul Raffaele continues in the *Smithsonian* magazine: "The locals [didn't] know where the foreigners' endless supplies [came] from and so suspect[ed] they were summoned by magic, sent from the spirit world" (2006).

Keep in mind that by many definitions of "well-being," including Maslow's, the case can be made the ni-Vanuatu were doing very well, thank you very much, prior to the Americans' wartime arrival. The ni-Vanuatu's physiological and safety needs were being met through fairly ready access to abundant and sustainable food, shelter and clothing. Their family and spiritual structure provided them with great opportunities to pursue Maslow's loftier goals of love, belonging, esteem and self-actualization. Therefore, if we define "affluence" in its non-monetary true meaning of "flowing in abundance," then when the American soldiers first arrived on Vanuatu soil, they were encountering a truly affluent indigenous people. Anthropologist Kirk Huffman, who lived in Vanuatu for almost twenty years, offers: "'Real' poverty [in Vanuatu] is almost non-existent in even the most remote areas, as everyone has land, food and culture" (2009).

When the war ended, so too did visits from the foreigners, who packed up their radios, TVs, trucks, boats, watches, iceboxes, medicine and Coca-Cola and headed home. In the aftermath of what had seemed to be magical material splendor, the ni-Vanuatu, as affluent as they had been prior to the soldiers' arrival, couldn't help but want them to come back. This is the context in which what some anthropologists call "cargo cults" came into being.

Anna Naupa and Sara Lightner highlight, in their "Histri Blong Yumi Long Vanuatu" (Our History of Vanuatu) series, that the term "cargo cult" has been held up as pejorative by some. Naupa and Lightner also note that some of the social movements associated with cargo cults began well before the arrival of the soldiers and their war-related supplies. The cargo cult movements often contained elements of protest that went beyond the worship of "cargo," such as the people's struggle against missionary rule. Indeed, Raffaele quotes one chief as saying that an important outcome of the soldiers' visit was "to help us get back our traditional customs, our kava drinking, our dancing, because the missionaries and colonial government were deliberately destroying our culture."

That said, it does seem to be the case that when the soldiers arrived with their wondrous possessions, many ni-Vanuatu were profoundly affected. Indeed, it is not surprising that some people assumed that magic and the spirit world were related to the arrival of these people and their things and that by tapping into these magical, spiritual realms, the cargo might return. Over the years, as efforts to encourage return failed, almost all of the cargo cults disappeared, but one survives on the Vanuatu island of Tanna. Some of Tanna's cargo cult activities focus on the more traditional forms of spirituality such as prayer and song, but the more striking activities involve ongoing attempts to recreate the circumstances under which the cargo first arrived on military planes and ships. As such, the Tannese people have maintained cleared swaths of land to serve as runways for the planes that will, hopefully, one day reappear, and they have constructed piers to accommodate large ships.

Such beliefs and activities seem quaint and naive from a distance, but there is a fundamental logic about the insecurity created when one's life as compared to another's *seems* not to be as good. I say *seems* not to be as good because what cargo cults highlight is that while it's easy to see when others have shiny new stuff and you do not, it is harder to see when you have the kind of affluence that is not contained in shiny new stuff. So, when the American military came to visit with their material riches, it is not surprising, in fact one might argue that it's downright logical, that many of the ni-Vanuatu were left longing for those magical items after the soldiers left.

This brings me back to stories. Humans are social creatures, and we make sense of our existence through stories. It took the unprecedented arrival of strangers from a strange land to shift the cargo cultists' understanding of their world and themselves. Here in North America, it is hard to imagine what a similarly radical parallel event would even look like (aliens arriving in their humungous spaceship is about as close as I can conjure), but that doesn't mean that we don't have our own opportunities for social comparison and our own cargo cults. While it is true that television's arrival wasn't quite as dramatic as glitter-draped troops arriving on our shores, what television lacks in dramatic entry, it has more than made up for in huge and consistent popular appeal.

In the decades since television's arrival, the medium has provided us with endless opportunities for social comparison and insecurity. What the comparisons have told us, both implicitly and often extremely explicitly, is that we can't possibly be happy the way we are — in those pants, without this makeup, living in that place, surrounded by those people — but we can be happier if we consume differently and consume more. Materialism can one day bring us happiness, television tells us. That is our cargo cult.

We don't keep mowing swaths of land for the planes that will come back "one day." No, our cargo cult involves the restless consumption of products for the increase in happiness that will come "one day." This too is logical. If we are endlessly told that we are incomplete and that stuff can complete us, and we are endlessly shown others for whom stuff has seemingly provided their "completion," then we learn: consume.

Are We There Yet?

Not only will our consumption make us happy and complete us as individuals, but we are also told that along the way we are performing an important role in our consumer society. As George Bush made clear in his encouragement of visiting Walt Disney World (September 27, 2001, speech) and shopping (November 8, 2001, speech) as two of the ways that Americans could defy the September 11, 2001, World Trade Center terrorist attack, consumption is part of our patriotic duty. Indeed, there are clear and well-publicized measures of whether we are performing this modern-day duty adequately. No such measure is more widely hailed than the gross domestic product, or GDP. Economist Peter Victor points out: "In a world that seems overflowing with statistics, the one that is most highly favoured over all others as a measure of progress is GDP" (2008: 9).

When one wishes to articulate "economic health," or "progress" (two concepts which, Victor points out, have fairly recently become synonymous), growth-of-the-market-based measures such as the GDP are where so many political and financial leaders turn. Economists like Victor are the first to point out that the GDP is not supposed to be a measure of anything more, or less, than total economic activity (total market value of goods and services) in a given area (usually a country) in a given period of time (often a calendar year). But the reality is that the GDP has — essentially since the Second World War — also been associated with the well-being of a country's inhabitants.

Victor cautions, however: "If we are to understand progress to mean an improvement in well being then GDP is a poor measure" (2008: 9). Victor offers several reasons for this. First, he highlights that GDP increases when expenditures are made because quality of life is *decreasing* — for example, when pollution gets so bad that abatement technologies have to be purchased, or when there have been extreme weather events and an area has to be rebuilt. Second, Victor points out that there are many aspects of our lives (often aspects that give our lives real meaning) that don't get calculated into the GDP — volunteer work, looking after our children, leisure time. A third element that Victor discusses is the fact that GDP doesn't address the distribution of the goods and services. In other words, GDP doesn't distinguish whether goods and services are spread amongst many or concentrated in few hands (the latter increasingly being the case).

In light of the shortcomings of the GDP, various alternative measures have been proposed in the past few years. These measures seek to move into a more nuanced measure of broad-based and genuine well-being. Perhaps the best known of these alternative measures is the Gross National Happiness (GNH) index, developed in the small South Asian country of Bhutan. From December 2007 to March 2008, the Bhutanese government undertook extensive interviews with the country's people in an attempt to articulate what the dimensions of the GNH should be. As a result of this rigorous undertaking, the following nine dimensions of the index were developed (each with several indicators): psychological well-being, time use, community vitality, culture, health, education, environmental diversity, living standard and governance.

The actual details of how each dimension is calculated and then compared across regions is complex and well-articulated on the Center for Bhutan Studies' Gross National Happiness website <grossnationalhappiness.com>. The key point is that the leadership and people of Bhutan have explored alternatives to what remains the default measure of a country's well-being, the GDP. As I alluded to above, the GNH index is in good company. Alternative measures such as the Genuine Progress Indicator (GPI), Human Development Index (HDI), Index of Sustainable Economic Welfare (ISEW) and Happy Planet Index (HPI) take into consideration indicators such as life expectancy, education, income distribution, household and volunteer work, crime, pollution, environmental capital and waste in the calculation of national well-being.

Let me highlight one of these alternative measures. The HPI was developed by the New Economic Foundation (NEF), an organization founded in 1986 which calls itself "an independent think-and-do tank." The HPI draws on E.F. Schumacher's famous book *Small Is Beautiful: Economics as if People Mattered*, saying that they "believe in economics as if people and the planet mattered." The goal of the HPI is to determine how efficiently the Earth's resources can be put to use in ensuring happy lives. Or, more succinctly and formulaically, the HPI calculates the average number of years of a happy life per unit of planetary resources consumed.

In 2006, the NEF formula was used for the first time to calculate the HPI for 178 countries. By combining the three indices of environmental footprint, life expectancy and happiness, it deemed Vanuatu the happiest country on the planet. Using the same three criteria, the United States was ranked 150th. By contrast, according to the traditional measure of GDP applied by the World Bank, in 2006, Vanuatu ranked near the bottom, at 207 of 233 countries, while the United States was number one. In 2009 and 2012, NEF undertook a second and third round of HPI calculations, and Costa Rica had the highest score, while the World Bank placed Costa Rica eighty-first in 2009 and eighty-fourth in 2011 (the United States remained at the top).

While developing alternative ways of understanding, and measuring, well-being is a crucial exercise, I certainly do not wish to make light of people's material wants and needs. About two-thirds of Vanuatu's population does not have access to electricity, let alone things that most North Americans (including myself) take for granted, such as indoor plumbing, hot water, a comfortable mattress and a huge range of food all year round. If you asked the ni-Vanuatu if they wanted these things, most would likely say yes. But as the NEF researchers highlight in their Happy Planet Index, the people of Vanuatu have ready access to incredibly rich natural capital (including plentiful local food all year); they experience excellent levels of democracy; they can expect to live a reasonably long life; and they express a high level of life satisfaction. That last part is important to note, because while those of us living in more affluent parts of the world may have access to more creature comforts, we are not happier.

It is easy to see why we would be perplexed that the endless-growth mantra of materialism has not made us happy. As Robert Lane asks in the introduction to his book *The Loss of Happiness in Market Democracies*: "How to account for this combination of growing unhappiness and depression, interpersonal and institutional distrust, and weakened companionship in advanced market democracies, in which people are, with important exceptions, reasonably well off?" (2000: 4). Increasing our flow of money and consuming stuff in ever greater quantities is what we've been endlessly encouraged to do. Yet somewhere along the way, the connection between our stuff and our well-being got lost.

Well-Being that Comes from "Enough"

You may have realized that there is an equation here. On one side of the equation there are people who lack well-being because they are materialistic, and on the other side of the equation there are people who are content because they lack materialism. Kasser and Ryan (1996) call the values that foster well-being "intrinsic values" — values that can be divided into three main categories (self-acceptance/personal growth, relatedness/intimacy, community feeling/helpfulness) and measured based on degree of agreement with the following statements:

> Self-acceptance/Personal growth
> I choose what I do, instead of being pushed along by life.
> I follow my interests and curiosity where they take me.

> Relatedness/Intimacy
> I express my love for special people.
> I have a committed, intimate relationship.

> Community feeling/Helpfulness
> The things I do make other people's lives better.
> I help the world become a better place.

When these values of self-acceptance/personal growth, relatedness/intimacy and community feeling/helpfulness are strongly present in our lives, we achieve a kind of contentment and happiness that psychology professor Mihaly Csikszentmihalyi describes as "a condition that must be prepared for, cultivated, and defended privately by each person.... It is by being fully involved with every detail of our lives, whether good or bad, that we find happiness.... It is a circuitous path that begins with achieving control over the contents of our consciousness" (2008: 2).

But so many of our lives are far from being "fully involved," and so many of us do not feel in "control of our consciousness." We are too busy working or going to school, sleeping and, yes, watching television. What we are cultivating tends to have much less to do with happiness and achieving control over the contents of our consciousness and much more to do with television cultivating the longing for more stuff. Csikszentmihalyi also ponders this:

> Why is it that, despite having achieved previously undreamed-of miracles of progress, we seem more helpless in facing life than our

less [materially] privileged ancestors were? The answer seems clear: while humankind collectively has increased its material powers a thousandfold, it has not advanced very far in terms of improving the content of experience. (2008: 16)

If you found that you were unable to offer "strongly agree" to many of the above intrinsic values statements, you are in good company. You are one of the legions of in-debt, overworked, unhealthy and unhappy people living in the affluent, high-GDP parts of the world. Under other global circumstances, the links between our propensity to watch hours and hours of television every day and television's propensity to instill materialistic values would be merely unfortunate. Life's short. If you can't get yourself off the couch, we might conclude, it's your problem.

At this moment in history, however, on this planet, our endless-growth economic model and the television-induced materialism that fosters it isn't personal and it isn't merely "unfortunate": it is global and it is cataclysmic. The lessons of endless insecurity and the materialistic salves that television have taught us haven't just resulted in malaise. No, our materialism is, as Schor poetically offers, "hurtling [us] toward an ecological precipice of unfathomable dimensions" (2010: 9). It is to a view from that precipice that I take you next.

4. ENVIRONMENT

Watching a whole lot of television is a personal decision, right? Coveting a whole lot of stuff in the belief that the stuff will make us better looking, cooler and happier, well, that's personal too, right? But what if television not only affects how materialistic we are, and what if that materialism not only makes us unhappy and unhealthy — what if television is also facilitating the destruction of the planet? What if the very foundations of life as we know it are crumbling under the weight of human consumption? Then we should take action, right? Then it would be correct to call for, even demand, drastic change, no? My task in this chapter is to convince you that television is killing the planet and convince you that we must demand change.

Environmental Crisis

A colleague of mine once joked that you haven't really written a paper about the environmental movement until you've mentioned that it all started with Rachel Carson's book *Silent Spring*. There are, of course, innumerable ways to tell the tale of modern-day environmentalism, but *Silent Spring* is a pretty good way to start. Published in 1962, *Silent Spring* offered a cautionary tale about the impacts of widespread chemical use to control weeds and insects. Carson's book was hugely popular (a bestseller when it was first published), hugely important (it now sits on several lists of the most important books of the twentieth century — including *Discover Magazine*'s top twenty science books and the conservative *National Review*'s top hundred non-fiction books) — and hugely hated (garnering scathing critiques and death threats). Even now, over forty years after death from breast cancer, there are those who battle her and her message.

What made Carson's work both terrific and terrifically threatening was not just that she deigned to interfere with the chemical industry's "business as usual," but also that she did so by proposing that what we do to the Earth, we do to ourselves. "The question" Carson wrote, "is whether any civilization can wage relentless war on life without destroying itself, and without losing the right to be called civilized" (1962: 99). If Carson were alive today, she would see clear evidence that this war on life has continued, relentlessly, and that we are, in fact, destroying ourselves.

One of the places that Carson would find strong scientific evidence of such destruction is the Millennium Ecosystem Assessment, established in 2001 by United Nations Secretary General Kofi Annan. As Annan states in the preface, they wanted to assess "consequences of ecosystem change for human well-being and the scientific basis for actions needed to enhance the conservation and sustainable use of those systems" (Millennium Ecosystem Assessment 2005). Carson certainly would have found the Assessment's findings disturbing, but she may have been comforted by the fact that these were not lone voices, like hers had been, but an entire community of scientists speaking about environmental devastation.

In fact, when Annan convened the Millennium Ecosystem Assessment, he brought together over 1,300 scientists and social scientists — the "largest group ever assembled to assess knowledge in this area" — and asked them to reach a *consensus* about the state of the world's ecosystems. Based on the evidence from the extensive analysis undertaken by this large group of scientific experts from 2001–2005, and with a $24US million budget, the main conclusion of the Assessment was as follows: "Human activity is putting such a strain on the natural function of Earth that the ability of the planet's ecosystems to sustain future generations can no longer be taken for granted." Let me drive this point home. Well over a thousand pre-eminent scientists come together to explore the state of the world's ecosystems, the largest group of this kind ever assembled to explore such a topic, and what is the point of full agreement at which they arrive? That humans are straining the Earth so greatly that the Earth may not be able to *sustain future generations*.

Other findings from the Millennium Ecosystem Assessment support Carson's notion that we are waging war on the Earth and ourselves. For example, the Assessment states that between 150 and 200 species become extinct every twenty-four hours, making this the greatest rate of species extinction in the last 65 million years. Economist Juliet Schor expands on this:

> We are in the midst of what biologists refer to as the sixth mass extinction. The last one happened 65 million years ago, with the loss of the dinosaurs. Among birds and mammals, extinction is occurring at a hundred to a thousand times the natural rates. (2010: 57)

In their article on mass extinction and amphibians, scientists David Wake and Vance Vredenburg conclude: "A primary message from the amphibians, other organisms, and environments, such as the oceans, is that little time remains to stave off mass extinctions if it is possible at all" (2008: 11472). Schor, and others, call this new reality of the environment not a crisis but "ecocide" —the mass killing of the planet's ecology.

While the reality of mass extinction and ecocide would have greatly saddened Carson, she would not have been surprised. The environmental phenomenon that might have been more surprising to Carson, both positively and negatively, is climate change. As with the Millennium Ecosystem Assessment, Carson would have been pleasantly surprised in the coming together of the world's scientists to study climate change. The Intergovernmental Panel on Climate Change, or IPCC, was created in 1988 by the World Meteorological Organization and the United Nations Environment Programme. According to economist Shardul Agrawala in his article "Context and Early Origins of the Intergovernmental Panel on Climate Change" (1998), what was noteworthy from the start about the IPCC was that it sought to combine quality scientific assessment with democratic consensus.

The IPCC now has 194 participating countries and thousands of scientists who have written a large number of publications including four assessments — in 1990, 1995, 2001 and 2007. In the first of these assessments, the consensus resulting from the IPCC's Science of Climate Change Working Group (Working Group Number 1) reads in part:

> We are certain of the following: emissions resulting from human activities are substantially increasing the atmospheric concentrations of the greenhouse gasses carbon dioxide, methane, chlorofluorocarbons (CFCs) and nitrous oxide. These increases will enhance the greenhouse effect, resulting in additional warming of the Earth's surface. (IPCC 1990)

One aspect of climate change that might have surprised Carson is its "upwardness." Carson's focus, and until recently most environmentalists' focus, was towards the ground. We spray, pour, gouge, clearcut and otherwise degrade the land on which we stand (or if there is concern about the air, such concern was on the air that we breathe). Climate change introduced us to a somewhat new concept in environmental degradation: what we release into the skies may not just make us cough, it may be what destroys us. Carson might also have found climate change surprising in that it is not caused by the intentional or explicit desire to alter the environment (as is the case with pesticides and herbicides), but rather as a result of day-to-day living and "business as usual." The cause of climate change can be found in the very foundations of our modern lives — heating and cooling our homes, manufacturing our stuff, driving our cars, producing and eating our meat. Carson would have been struck by the fact that our seemingly mundane activities have brought us as close as humans have come to truly waging environmental war on ourselves.

When *Silent Spring* was published in 1962, its message was both startling and stark. "How could intelligent beings seek to control a few unwanted species," Carson asked us, "by a method that contaminated the entire environment and brought the threat of disease and death even to their own kind?" (1962: 8). Now, all of these years later, the message is no longer startling. We've heard a lot about environmental degradation and climate change. The environment and all things "green" are discussed, advertised, debated and sold to us everywhere we look. And we're doing our part by undertaking good environmental actions. We put out our recycling bins and sometimes we buy the "green" alternative. We have even made changes to the laws regarding the pesticide use about which Carson was so profoundly concerned. For example, in April 2008, Ontario, Canada, introduced Bill 64, the *Cosmetic Pesticides Ban Act*, which, as the name suggests, bans the sale and use of pesticides for cosmetic purposes.

Species loss, however, continues unabated, and climate change looms larger than ever. Carson's explanation in 1962 works just as well today:

> The road we have long been traveling is deceptively easy, a smooth superhighway on which we progress with great speed, but at its end lies disaster. The other fork of the road — the one "less traveled by" — offers our last, our only chance to reach a destination that assures the preservation of our earth. (1962: 277)

Of course, the question all these years later is *why* we don't choose another road.

Haven't We Been Here Before?

One fundamental reason for not getting off of our easy, smooth superhighway is that we don't, at some gut level, believe that disaster looms at its end. Let me start by saying that I think skepticism makes sense. All of a sudden we're being bombarded by all of this environmental crisis "doom and gloom" talk, and if you're paying attention at all, you have every reason to ask a fundamental question: *Haven't we been here before?* Throughout history there have been claims that, essentially, the sky is falling. The details have varied. In the late nineteenth century, when horses ruled our travel options, concern was for the apocalyptic implications of their manure. In his article "The Great Horse Manure Crisis of 1894," Stephen Davies explains the manure fears:

> The larger and richer that cities became, the more horses they needed to function. The more horses, the more manure... one writer estimated that in 50 years every street in London would be buried under nine feet of manure. Moreover, all these horses had to be stabled, which used up ever-larger areas of increasingly valuable land. And as the number of horses grew, ever-more land had to be devoted to producing hay to feed them (rather than producing food for people), and this had to be brought into cities and distributed by horse-drawn vehicles. It seemed that urban civilization was doomed. (2004)

Clearly horse manure did not mean the end of civilization; looking back we are struck by the quaintness of such concerns.

Other fears about civilization's impending doom were less trivial. During times of large-scale disease outbreaks people had very real reasons to believe the end was near. Indeed in *The Great Pestilence AD 1348–1349: Now Commonly Known as the Black Death*, Francis Aidan Gasquet highlights some of characteristics of the bubonic plague, "1) Gangrenous inflammation of the throat and lungs; 2) Violent pains in the region of the chest; 3) The vomiting and spitting of blood; and 4) The pestilential odour coming from bodies and breath of the sick" (1985: 7). For millions of people, this is what characterized the end of life. As Suzanne Alchon offers in *A Pest in the Land: New World Epidemics in a Global Perspective*, the plague decreased the population of some European villages and cities by as much as 80 percent. Now *that* would feel like the end of the world. For people living at that time, the reality that there was a crisis was never in doubt, the crisis was *everywhere* and the end of the world seemed a certainty.

Disease did not, however, wipe out humanity. Some, like Paul Ehrlich, proposed that the end would come not due to annihilation from disease but the antithesis: humanity would meet its demise from our very heartiness and overabundance. Ehrlich's book *The Population Bomb*, published in 1968, conveyed various dire predictions for what would happen in the face of continuing population expansion. The essence of those predictions was that the birth rate must be brought into balance with the death rate, and even if we managed to do that (with great effort), huge numbers of people would die of starvation. Ehrlich surmised that there just wouldn't be enough resources, especially food, to go around and cautioned that without restricting the birth rate, "mankind will breed itself into oblivion." While Ehrlich continues to maintain that he was correct in signalling concern for the stress an increasing global population will place on the planet, by all accounts his dire predictions for humanity's demise (at least in the relative short term) were wrong.

Demise in the short term did seem like a real possibility during the Cold War of the 1980s, when stockpiles of nuclear weapons and tense relations between the Soviet Union and the United States made it feel like we were often teetering on the brink of large-scale destruction, even annihilation. Of course we continue to be a species that has at our fingertips the unprecedented and very real ability to quickly annihilate ourselves (and most of the life with whom we share the planet). That said, world leaders have, with the exception of two nuclear bombs dropped on Japan during the Second World War, been mercifully reluctant to make use of such weaponry.

Other more recent end-of-the-world-type scares have, for most of us, seemed much less serious than the prospect of nuclear holocaust or annihilation through population explosion or plagues. For example, in the lead-up to the shift from the year 1999 to the year 2000, there were "Y2K" fears that a large-scale crisis would occur as our technology-dependent world became unable to function when the calendar clicked over from the twentieth to the twenty-first century (the idea being that our newly wired world would fail as the computers were unable to "make sense of" the change from 1999 to 2000). Then there are the periodic apocalyptic warnings with biblical origins. For example, Harold Camping, founder of the Family Radio Network, has been predicting Jesus' second coming and thus the destruction of the world. Each time Camping has incorrectly predicted Jesus' arrival, a new arrival date is set — May 21, 1988; September 7, 1994; May 21, 2011 and October 21, 2011 — and each time there is a new date, there are many followers who prepare.

Perhaps the most tragic end-of-the-world-prediction scenarios are those where the handful of faithful take their own lives in preparation. One of the most notorious examples is that of the Heaven's Gate suicides in March 1997, when thirty-nine people killed themselves in San Diego, California, to avoid what their leader, Marshall Applewhite, claimed was the inevitable destruction of Earth by the Hale-Bopp comet.

I highlight these crises because they are all well known, and while some were truly frightening and devastating, the world has chugged along. We stopped using horses for transportation; the plagues ended (it would seem due to some combination of the strong being able to fight the diseases and an increasing awareness of the importance of hygiene); for now there has not been a Second Coming nor a catastrophic meteor; we look somewhat sheepishly back on our Y2K concerns; and the planet's population is seven billion and counting.

That's one reason we don't believe that the smooth paved road we're on ends in disaster: we've been warned before about impending doom and we're still here. There's another reason we don't leave the smooth paved road we're on: everything *seems* fine.

Where's the Environmental Crisis in My Life?

From the perspective of your life, do you think there is an environmental crisis? Are *you* experiencing an environmental crisis? I don't mean the last time you watched the news or read a newspaper, but this morning, when you got up for work? Did it seem like there was an environmental crisis underway? Probably not. The reason we don't see the crisis is twofold: we're too busy and we don't know what we're looking for.

Like me, many of you probably got up this morning and made the observation that in spite of having slept, you were still tired (these thoughts may have set in motion a plan to find some coffee). If you have kids, you might have noticed that it can be incredibly hard to organize them and get them out the door to go anywhere. You may have realized, as you raced around trying to leave your house, that you still haven't gotten to that task you've been meaning to do (the wobbly chair, the paint chipping on part of the wall, the leak under the sink...). If you're a student, perhaps you are stressing about the readings you didn't get to. You may have also realized that there's nothing in the fridge you feel like making for dinner (you'll have to pick up some prepared food on the way home). But chances are, when you walked out of your house this morning you did not think to yourself, "Now *this* is an environmental crisis."

Scholars have analyzed why we don't walk out of the house in the morning and think "environmental crisis." One key reason is the awareness we traded in when we became the "modern humans" we are today. In *The Reenchantment of the World,* Morris Berman proposes that our disengagement with the natural world began with the Scientific Revolution in the seventeenth century. Prior to that we experienced the Earth as being wholly alive and a place with which we were fundamentally connected. Berman contends that this holistic mentality "made virtually no distinction between subjective thought processes and what we call external phenomena.... This notion, that subject and object, self and other, man and environment, are ultimately identical, is the holistic world view" (1989: 65). The Scientific Revolution required that we separate ourselves from that which we study. We were no longer fundamentally connected to the Earth; we became objective observers and measurers. Centuries later, when we walk out of our home, therefore, what is going on in our minds and bodies is separate from what is outside of us — we are separate from the environment.

In order to take on the scientific worldview we had to leave behind this "holistic" or "animistic" understanding that we and the Earth are intimately and inseparably connected. "For what was ultimately created by the shift from animism to mechanism was not merely a new science," offers Berman, "but a new personality to go with it" (1989: 105). What this "new personality" involved was understanding that the world around us was no more, or less, than substances to be harnessed in the name of science and materialistic "progress." Humanity was no longer part of the "ageless rhythm of ecology," but rather the "controller and manager" of the world such that we could try to satiate our growing need for consumption, what Berman terms our "fetishism of commodities" (34, citing Brown's *Love's Body*). Thus, Berman would say that it is difficult for us to recognize environmental crisis as we go about our lives, because what we see when we look at the world around us are the manipulated landscapes and commodities with which we have always lived and which have been removed from their origins.

In *The Death of Nature: Women, Ecology and the Scientific Revolution,* Carolyn Merchant also writes about the historic origins of our disengagement from the environment.

The world we have lost was organic. From the obscure origins of our species, human beings have lived in daily, immediate, organic relation with the natural order for their sustenance. In 1500, the daily interaction with nature was still structured for most Europeans, as it was for other peoples, by close-knit, cooperative organic communities.... As European cities grew and forested areas became more remote, as fens were drained and geometric patterns of channels imposed on the landscape, as large powerful waterwheels, furnaces, forges, cranes and treadmills began increasingly to dominate the work environment, more and more people began to experience nature as altered and manipulated by machine technology. A slow but unidirectional alienation from the immediate daily organic relationship that had formed the basis of human experience from earliest times was occurring. (1989: 68)

Throughout most of human history we would have recognized an environmental crisis because we intimately knew our environment. In recent history we began to manipulate the environment to suit our scientific and social needs, and our relationship has become one of mastery over, rather than synergy with, the environment. As a result of this change, it has become extremely difficult for us to see that our environment is being destroyed.

Our alienation from the environment's "deep constant rhythms" is a theme that Bill McKibben similarly explores in his book *The End of Nature*. McKibben articulates, and mourns, that:

We can no longer imagine that we are part of something larger than ourselves — that is what all this boils down to... now *we* make that world, affect its every operation (except a few — the alteration of day and night, the spin and wobble and path of the planet, the most elementary geologic and platonic processes). (1989: 84)

As McKibben points out, it can be easy to mistake our affecting of the environment for our independence from it and our control of it. It is easy to believe that if our actions caused these things to happen to the environment, our actions can also undo what we have done. No need to worry.

So, you see, there are good historical reasons for why you don't step outside of your house and think, "environmental crisis." We're too disconnected and/or too arrogant to know better. Added to this is the fact that if we want to believe that there's no need for environmental alarm, the successes of our "modern society" are in many ways more tangible and therefore easier to find than the indicators of a threatened planet. The average North American can expect to live longer than ever before. When a woman in North American has a baby she can reasonably assume that both she and her baby will survive. The vast majority of North Americans have indoor plumbing that includes the opportunity to have relatively clean drinking water, and even hot bathing water, at the simple turn of a knob. Turning a knob is also all we need to do if we want to warm or cool our home or cook or cool our food. In fact, amid record heat in the summers of 2011 and 2012, air conditioning use resulted in some of the highest electricity consumption on record throughout North America. Stan Cox, senior scientist at the Land Institute and author of *Losing Our Cool* (2010), points out that between 1993 and 2005 the energy needed to run our residential air conditioners doubled, and air conditioning-related energy use continues to grow by leaps and bounds (2012). Our ancestors may have had an organic relationship with the environment and may have understood the ageless rhythms of ecology, but it is easy to argue that our environmental alienation is a small price to pay for longevity, safe childbearing and predictable access to comforts that, until recently, would have been unimaginable.

Life, in general, is pretty good for us. If there are warning indicators within the environment, we're too disconnected from them, and as such too ignorant of their manifestations, to notice or really care. There are some people, those pesky activists for example, who have been trying to get our attention about environmental issues, but these people have been shouting about environmental concerns for decades now, and, really, our lives have pretty much chugged along the same as ever (with perhaps a few nods to being "green").

Or have they? It is true that we are no longer aware of the environment in the subtle and organically connected ways we once were. It is also true that because of this disconnection it makes sense that we would divest responsibility for being aware to "the experts" — those who undertook the Millennium Ecosystem Assessment and belong to the IPCC, for example — and let them tell us about the state of the environment. I think, however, that we know more about fundamental environmental changes than we realize. My guess is that there are not entirely "scientific" indicators of environmental distress that *are* available to us and to which we do, indeed, pay attention.

At the beginning of my environmental communication class, I ask students to share their "environmental autobiography" with the class. In this autobiography (a variation on the television autobiography I talked about in the Television chapter), students are encouraged to talk about what their relationship with the natural environment has been like throughout their lives. Of course, there is a great range in the kinds of relationships students have with the environment. Some spent their childhoods in the outdoors camping, exploring and adventuring — and loving every minute. Others never placed a foot outside except to get from point A to point B, maybe attend the occasional barbeque in the back yard, and they feared every bug they encountered, hated the glaring burning sun and couldn't wait to return to the sanity of the indoors and electricity. One of the common themes in these students' tales, however, wherever they fall on the love-or-hate-the-environment continuum, is the experience of some kind of environmental loss. Students will talk of spaces that were once green, once had trees, once had animals, and are now gone... for subdivisions... grocery stores... highways.

I had this experience. My earliest school memories involve romping around a forest that surrounded my school. It was an alternative school that placed value on outdoor play, and so my memories of endless hours of exploring the forest and climbing the trees are not exaggerated. Many years later when my brother and I returned to the school, the forest was gone. In its stead was a highway. There remained a few token trees (my brother and I clambered onto one and took pictures) and groomed grass, but the highway ramp had literally overtaken the view behind the school. You probably have a similar story — perhaps many. While such stories are, in the language of researchers, "anecdotal" and not of great value when deciding whether to declare that the road we are on will end in disaster, when such personal narratives are added to the massive formal research undertaken by scientists in groups such as the Millennium Ecosystem Assessment and the IPCC, then we know that we need not solely rely upon and trust the experts. We can add our personal histories and experiences to their expertise.

Why don't we leave the superhighway we're on, the one that Carson seems to have correctly predicted is leading us to disaster, the one that we can see from our personal experience has led to environmental degradation? I propose that part of the answer is the "crying wolf" phenomenon: we've heard before that the sky is falling but it hasn't happened yet, so many of us are now skeptical. Another part of the answer is that while we may be able to draw on examples of environmental loss in our lives, day-to-day we do not experience an environmental crisis. Underpinning all of this, what has most successfully propped up our collective denial is the effective storytelling from our favourite storyteller: television. Television's unrelenting and omnipresent stories demand, in the nicest possible way, that we stay on the superhighway.

Environmental and Television Autobiographies

As you now know, I'm a fan of using autobiographies — environmental autobiographies, television autobiographies — as a way to tap into personal relationships with these topics. In order to propose what I believe exists at the intersection of television and the environment, let me start by offering a few parts of my own television and environmental autobiographies.

My television autobiography goes something like this: similar to many young people growing up in the latter half of the twentieth century, I watched television. My memories of television include watching cartoons on weekend mornings with my brother and in the evenings sit-coms, drama and variety shows. Three aspects of that early television viewing stand out for me. First, even then I knew I was affected by television (feeling groggy after viewing and often drawing upon unpleasant images and stories in the scary darkness of my bedroom). Second, I knew I should be wary of television because my parents limited viewing time (long before the research so explicitly suggested that television viewing should have limits). Third, I was always aware that television was, at its core, a business. This latter fact was driven home for me because my father was in the radio business. A radio was always on at our house, and my dad's obsession with ratings was a constant reminder that mediated communication is all about the relationship between creating audiences and selling them to advertisers. Success, almost without exception, is measured in one way: size of audience (the larger the audience, the more advertisers are willing to pay).

My environmental autobiography goes something like this: I grew up barefoot and climbing trees. I was born in Florida, and the warmth of that part of the world, combined with the relative lack of sprawl, meant that it was easy to be barefoot and easy to find trees to climb, a river to navigate and dirt to dig. When Richard Louv, author of the book *Last Child in the Woods* (2005), notes that boundaries for kids used to be measured by blocks or miles (as opposed to home and backyard restricted kids today), it rings true for me. I roamed endlessly with my friends. My mother would have no idea where we were so she would use a bell to let us know it was dinnertime. That was the one rule: be within the range of the bell. When I was eight we moved to Canada, and my outdoor adventures came to include long, wondrous canoe trips in northern Ontario and then some rock climbing and kayaking in British Columbia. My love of being in the outdoors became a love of working for the outdoors, and I later spent many years volunteering and working for environmental organizations such as Pollution Probe and Greenpeace.

As an environmentalist, one of the most perplexing Rachel Carson-esque questions I kept returning to was how we, as a species, could have arrived at such a pathological relationship with that which sustains us. This is why when I first encountered the writings of people like Morris Berman, Carolyn Merchant and Bill McKibben (all quoted above), I was very excited. Their writings about our relationship with the environment helped to make logical that which seemed so entirely illogical. After all, we really are a strange and unique species; we stand apart from every other creature on the planet in our active destruction of our habitat. All other species explicitly respect their habitats through their actions (for example, not defecating in their nests and dens) or at least implicitly respect the balance of life through their intimate connection to the web of life (for example, ebbing and flowing in numbers as quantities of food and prey fluctuate).

In recent history we humans have been neither explicit nor implicit respecters of our habitat or the balance of life. This is why Ronald Wright's reference in *A Short History of Progress* to the way in which we live on this planet as a "suicide machine" (2004: 131) makes sense to me. How can a species that prides itself on its logic and rational thought (indeed a species that likes to propose that its logic and rational thought is what *distinguishes* it from other species) wind up living so recklessly? Wright offers:

> We tend to regard our age as an exception, and in many ways it is. But the parochialism of the present — the way our eyes follow the ball and not the game — is dangerous. Absorbed in the here and now, we lose sight of our course through time, forgetting to ask ourselves.... *Where are we going?* (2004: 109)

When I first went to graduate school, these were the kinds of questions that guided me: How did we get here? Where are we going? I was also struck by the fact that humans have only recently lived in the way that we do. As authors like Berman, McKibben and Merchant highlight, for millennia many humans lived such that their lives had a kind of balance, or rhythm, with the rest of the planet. There is always the danger of romanticizing, but it is fairly easy to argue that the planet's original people, or indigenous people, usually did a better job of living harmoniously with the Earth. Even when a seeming counter-example is held up, such as is done with the buffalo or bison jumps of the northern plains (where sometimes more than a thousand animals were stampeded off a cliff and then killed), research indicates that there was still balance. For example, in the book *Imaging Head-Smashed-In: Aboriginal Buffalo Hunting on the Northern Plains*, Jack Brink argues: "The hunting strategy and belief system of the ancient people of the Plains was clearly a successful one; it supported their way of life for thousands of years without leading to a decline in numbers of bison" (2008: 160). This too is part of my environmental autobiography: exploring the way in which other people have lived on the planet without destroying it.

Many indigenous peoples have had cultural practices that were explicitly about maintaining balance through respect for the non-human world. One practice that played a central role in how they understood their relationship with the environment was storytelling. As Carolyn Merchant points out, respect for the land was manifest in indigenous people understanding and celebrating the Earth as the "mother." Such a metaphor encouraged people to understand, fundamentally, how humans are reliant on the Earth for the same things that children need from their mother: sustenance, protection, meaning. Merchant writes:

> The earth, or geocosm, was universally viewed as a nurturing mother,

sensitive, alive and responsive to human action. The changes in imagery and attitudes relating to the earth were of enormous significance as the mechanization of nature proceeded. The nurturing earth would lose its function as a normative restraint as it changed to an inanimate dead physical system. (1989: 20–22)

Jerry Mander highlights in his book *In the Absence of the Sacred: The Failure of Technology & the Survival of the Indian Nations* that this practice of understanding the Earth as mother was common.

It is not only American Indians who use the phrase ["Mother Earth"]. So do Aborigines of the Australian desert, natives of the Pacific Islands, Indians of the Ecuadorian jungles, Inuit from Arctic Canada; in fact, I have yet to find a native group that does not speak of the planet as mother." (1991: 212)

How was this understanding of the Earth as mother conveyed? Stories. Thomas King writes in *The Truth About Stories: A Native Narrative*:

While the relationship that Native people have with the land certainly has a spiritual aspect to it, it is also a practical matter that balances respect with survival. It is an ethic that can be seen in the decisions and actions of a community and that is contained in the songs that Native people sing and the stories that they tell about the nature of the world and their place in it, about the webs of responsibilities that bind all things. (2003: 114)

I am not, however, as interested in how we once told stories as I am in how our current storyteller, television, shapes our relationship with the Earth.

Academic Field of Environmental Communication

I am certainly not the first researcher interested in the relationship between television's stories and how we understand our connection with the Earth. Some researchers have drawn upon the quantitative cultivation theory approach (explored in the Television chapter), and others, such as Jerry Mander and Bill McKibben, have tried to make sense of the relationship between television and the environment more qualitatively and critically. The qualitative and critical approaches have largely revolved around television's time displacement (when we watch hours of television each day this reduces the time we could be doing other things, like roam outdoors). Mander says: "[Sitting] in our dark rooms ingesting images of Borneo forests, we lose feeling even for the forests near our homes. While we watch Borneo forests, we are not experiencing neighborhood forests, *local* wilderness or even *local* parks" (1978:

280). Qualitative and critical analyses of television's storytelling limitations have also involved television's sensorial constraints. Mander gives this rather eloquent lament for television's limitations in telling the story of a marsh:

> Images and words about a marsh do not convey what a marsh is. You must actually sense and feel what a strange, rich, unique and *un*human environment it is. The ground is very odd, soft, sticky, wet and smelly. It is not attractive to most humans. The odor emanates from an interaction between the sometimes-stagnant pools and the plants that live in the mud in varying stages of growth and decay. If the wind is hot and strong, there can be a nearly maddening mixture of sweet and rotting odors. To grasp the logic and meaning of marsh life, the richest biological system on Earth, one needs to put one's hand into the mud, overturn it, discover the tiny life forms that abound. One needs to sit for long hours in it, feeling the ebbs and flows of the waters, the creatures and the winds. (1978: 279)

In many ways Bill McKibben's second book, *The Age of Missing Information* (1993), picks up where Jerry Mander left off. In McKibben's first book, *The End of Nature* (1989), exactly as the title suggests, the author laments the demise of nature. It is not that McKibben believes that there are no remaining places of "nature." Instead he proposes that while throughout most of human history the natural environment held a sacred and mystical place for humans, this sacredness and mysticism have been lost in the last century. McKibben argues that once everything can be associated with human activity, then the natural environment no longer has the potential to contain the unknown. The environment can no longer be spiritual. "The *meaning* of the wind, the sun, the rain — of nature — has already changed. Yes, the wind still blows — but no longer from some other sphere, some inhuman place" (1989: 48). How did this happen? How did we get to this place where humans have left their (often devastating) mark on every aspect of the environment?

McKibben claims that the change in our relationship with the environment has everything to do with the lives of a small number of people. "The way of life in one part of the world in one half-century is altering every inch and every hour of the globe" (1989: 46). But how did we *learn* this way of life? This way of thinking? McKibben is vague in explaining what happened. "Almost all of us intuitively hold this idea of infinite progress," he proposes, "having imbibed it with our infant formula through the sterilized rubber nipple" (1989: 154).

Interestingly, McKibben mentions television only once in *The End of Nature* — he makes a passing reference to his sense that while people might watch nature shows from time to time, mostly we watch fictional shows like *L.A. Law*. However, in *The Age of Missing Information* (1993), McKibben turns his attention entirely to television in order to explore its relationship with our loss of environmental connection. In fact, McKibben participates in a personal experiment that taps into the heart of what Mander's marsh comment and cultivation theory propose.

There are two parts to McKibben's experiment. First, he tapes twenty-four hours of the ninety-three cable television channels that are available to him in his Fairfax, Virginia, home. He then watches the tapes (over a thousand hours) and reflects on what he learns about the world. He concludes, not surprisingly, that television's content is commercial and vapid. He highlights as part of this conclusion the way in which television's stories take over our worldview. "I assumed unconsciously that the information that poured from the TV into my quite similar suburban world was all the information there was" (1993: 248).

The second part of McKibben's personal experiment involved spending twenty-four hours hiking and camping in the Adirondacks, then reflecting on the "lessons" learned from that experience. One of the key pieces of information his time in the mountains offered him is the following:

> Human beings — any one of us, and our species as a whole — are not all-important, not at the center of the world. That is the one essential piece of information, the one great secret, offered by any encounter with the woods or the mountains or the ocean or any wilderness or chunk of nature or patch of night sky. (1993: 228)

McKibben experienced what cultivation theory measures: that our understanding of the world is shaped by television (and the more television we watch, the more shaping that takes place). McKibben's experiences would lead us to predict that the more television we watch, the more likely we are to believe that humans are all-important, sitting at the centre of the world (and as you will see below, cultivation theorists have tested this proposition and found it to be true). But McKibben's personal experiment offers us another proposition: *the natural environment also teaches and cultivates us.* The more time we spend in nature, the more likely we are to access its lessons, such as: "Information about the physical limits of a finite world. About sufficiency and need, about proper scale and real time, about the sensual pleasure of exertion and exposure to the elements, about the human need for community and for solid, real skills" (1993: 236).

McKibben's lack of rigorous methodology and Mander's at times over-the-top rhetoric have made them easy for "serious scholars" to dismiss. Their conclusions about the role television plays in helping us understand the world around us are not, however, all that much different than the conclusions at which "serious scholars" of environmental communication have arrived. Simply put, "environmental communication" is any exploration that takes place at the intersection of communication (from two people chatting to millions watching the same television program) and the "natural environment" (or the Earth, nature, wilderness). In his book *Environmental Communication and the Public Sphere*, Robert Cox defines environmental communication as, "a study of the ways in which we communicate about the environment, the effects of this communication on our perceptions of both the environment and ourselves, and therefore, on our relationship with the natural world" (2009: 9).

How to mark the beginning of this field of inquiry and exploration is somewhat up for grabs. We could, once again, draw upon Rachel Carson's publication of *Silent Spring* in 1962 as some kind of starting point. Carson was certainly conscious of the role of environmental communication when she made reference to the *New York Times* raising awareness of the dangers of pesticides by issuing a warning (176), and portions of *Silent Spring* initially appeared as installments in that newspaper. Most significantly, *Silent Spring* is one of the, if not the, pre-eminent examples of environmental communication. All of which put Carson squarely within the category of environmental communication pioneer. That said, in the 1960s, the field of environmental communication was several decades away from taking off.

In an article I co-authored entitled "The Literature of Environmental Communication," we found one article published in 1948 that could be categorized as environmental communication. The field did not, however, really begin to blossom until the mid-1980s. In fact, in 1985 the number of environmental communication articles being published doubled from the previous year, and between the years 1985 and 1990, the average annual growth rate in published environmental communication research was 44 percent (Pleasant, Good, Shanahan and Cohen 2002).

Coinciding with these increases in environmental communication publishing in the mid-1980s and early 1990s were other notable milestones in the formation of environmental communication as a field of research. For example, in 1988 a panel called "The Discourse of Environmental Advocacy" took place at the Speech Communication Association's annual conference — marking what is possibly the first environmental communication panel at an academic conference (Senecah 2007). In 1991 the first Conference on Communication and the Environment (COCE) took place in Alta, Utah (COCE has taken place every other year since), and in 1993 the Environmental Communication Working Group was established as part of the Speech Communication Association. The field of environmental communication has grown quickly over the years since these early initiatives. Various conferences now have environmental communication divisions and several university departments focus on environmental communication. In 2007 *Environmental Communication: A Journal of Nature and Culture* was born, providing a focused publication venue for environmental communication researchers, and in the summer of 2011 the International Environmental Communication Association was formed.

Not surprisingly, within this burgeoning field there have been, and continue to be, myriad research approaches. In *Nature Stories: Depictions of the Environment and Their Effects* (1999) James Shanahan and Katherine McComas provide a detailed overview of environmental communication research at the intersection of mediated communication, especially television, and the environment. They divide their overview of the research into three categories: first, research that has explored content of environmental communication — looking at how television has portrayed the environment; second, research that has critically explored the social and cultural implications of television's stories about the environment; third, research into the effects of television's consumption on individuals' environmental attitudes, values and beliefs.

As the authors highlight, while the environmental communication research that has focused on content is interesting, such studies are limited in that content studies are unable to tell us about the *effects* of the contents on the audience. Indeed, "these [content] studies point to the need... for work in the effects area to understand how these complex content patterns are internalized by mass media audiences" (Shanahan and McComas 1999: 35). Similarly, while critical analyses of environmental communication can be both interesting and valuable, such analyses often provide broad, sweeping contexts for exploration, but not a whole lot of detail about causation (what is affecting what and why). Shanahan and McComas propose, therefore, that cultivation theory is best suited to explore the causation questions that we have about the relationship between television and the environment. Cultivation research begins with a television content analysis (an important step, but not enough) and then proceeds to test what the effects of that content are for viewers.

If you recall from the Television chapter, cultivation theory proposes that those who watch more television are more likely to take on "television's understanding of the world" than those who watch less television. Over the years, the majority of cultivation research has been done on the relationship between levels of television viewing and how people think about violence and aggression in the real world. For example, the "mean world syndrome" is the term that cultivation theorists developed to explain the phenomenon that those who watch more television tend to think of the world as a meaner, scarier place than those who watch less television. The Materialism chapter explores research at the intersection of cultivation and materialism, which has found that, on average, those who watch greater quantities of television are more materialistic than those who watch less television.

In addition to these, and numerous other topics, cultivation theory has also been used to explore the relationship between television viewing and how people understand the natural environment. As I have also previously detailed in the Television chapter, cultivation research has three components — what television has to say about a particular topic (the content analysis), how heavy and light television viewers compare in their understanding of that topic (the cultivation analysis) and the cultural context in which television's messages are created (the industry analysis).

Environmental Cultivation: Content Analysis

What do content analyses tell us about television's portrayal of the environment? Some of the earliest content analyses of mediated communication for environmental content were done on magazines and other printed periodicals. In 1977, James Bowman and Kathryn Hanaford published their study of mass circulation magazines entitled "Mass Media and the Environment Since Earth Day." In their analysis of eight top circulation magazines — from January 1971 (the year of the first Earth Day) to December 1975 (following up on a similar study by David Rubin and David Sachs that analyzed magazine content in the 1960s) — Bowman and Hanaford concluded:

> The evidence presented here demonstrates that general circulation magazines give but minimal coverage to [the environment] one of the most significant issues in the last part of this century.... Since the general public receives much of its information from the media and will base many of its actions on that information, the extent of commitment on the part of the media seems critical. (1997: 164)

In 1981, James Bowman and T. Fuchs published the results of a similar study that set out to "gauge the history of contemporary environmental concern as reflected in leading periodicals" of the 1960s and 1970s. The authors' conclusion echoes the findings from the Bowman and Hanaford study:

> Evidence has been found that media executives see the crisis as real, but that it is relatively unnewsworthy. ... Mass circulation magazines must provide important information and continue to improve their reporting of the environmental crisis. (1981: 21)

As you now know, however, cultivation theory's focus is on television (and not magazine) content. It wasn't until about twenty years after these early environmental content studies that the first analyses of environmental content on television were published. In 1991, Robin Peterson published her analysis of "ecologically responsible" themes in television advertising. Peterson made use of a television sample of the programming between 8:00 a.m. and 12:00 p.m. on two randomly chosen days each month in the years 1979 and 1989.

The three major United States networks, one local station and five cable companies provided the programming. Peterson and her coding assistants defined an "ecologically responsible" commercial as: (1) a commercial which directly advocated an ecologically responsible theme (air pollution, water pollution, noise pollution, depletion of scarce resources, destruction of the landscape, population explosion); (2) a commercial which depicted an ecologically responsible participant in a favourable light; and (3) a commercial which favourably illustrated a goal of an ecologically responsible issue (224).

In her findings, Peterson calls the number of "ecologically responsible" advertisements "moderate" and points to a slight increase in the number of such advertisements from 1979 to 1989. She also highlights that while in 1979 the most popular ecological theme in the advertisements was depletion of scarce resources, in 1989 the most common themes were water and air pollution. While these content findings are interesting, the focus on advertisements makes the study somewhat different from cultivation content analyses, which usually focus on a broader sample of television content such as prime-time (e.g., between 7:00 pm–11:00 pm) fictional programming.

It was not until the 1990s that the first environmental cultivation content analyses were published. In a series of studies, James Shanahan became the person most associated with laying the foundations for environmental cultivation research. In his initial research, Shanahan randomly chose a week of programming in November 1991, January 1993 and February 1994 on three network affiliate stations (ABC, CBS, NBC) during prime-time. This represented 317 hours of content and 402 separate programs. Shanahan collected various kinds of information about the programming, but the focus of the content analysis was on how frequently, and in what ways, the environment was present. Shanahan looked for what he called "environmental episodes," which he defined as "any discrete portion of a program involving spoken words or physical action in which environmental issues were specifically implicated or discussed" (Shanahan and McComas 1999: 85). These episodes were then further coded for such things as topic (e.g., recycling, climate change, endangered species), who was involved with the episode and whether the episode was "concerned," "neutral" or "unconcerned" regarding the environment. It should be noted that just because a natural setting such as a lake or a mountain range was part of a scene, this did not in and of itself establish nature as a theme in a program. In order to be considered "a theme," nature or an environmental issue had to be part of the story (86).

What the researchers found did not surprise them and probably won't surprise you: nature/the environment was largely absent as a theme. In 80 percent of the programming the environment was completely absent, and in only 1.7 percent of the programming was the environment an "outstanding theme" (i.e., the focus of the program). "Compared to other themes such as entertainment, family, personal relationships, and financial success," Shanahan and McComas highlight that, "nature as a theme is very infrequent" (86).

In 2001, Katherine McComas, James Shanahan and Jessica Butler published the results of a study that built on the above research. In this study the sample was broadened to include the years 1991–1997, and the Fox news network was added to ABC, CBS and NBC for the year 1997. The content focus, however, was narrowed to include only prime-time entertainment/fictional and non-news programming (as is typical of cultivation research). Otherwise the approach was very similar to Shanahan's earlier television environmental content research and the findings were also similar: the environment is largely absent from television's programming. In the six-year sample of 410 hours of non-news entertainment and fictional television, the time allotted to the environment was approximately two hours and twenty-two minutes, or about half of 1 percent of the total hours that were watched (2001: 538).

The researchers also analyzed the content of those two hours and twenty-two minutes of environmental programming. For example, McComas and her colleagues found that television's environmental content tended to stand alone and wasn't integrated with the much more prominent programming themes like home, family and relationships. The researchers also found that those television characters who were involved with the environmental activities were most likely to be white, middle-aged, middle class and male. This is important because Gallup Poll results from the 1990s indicated that "real people" who favoured giving priority to environmental protection over economic growth were more likely to be non-white females, aged eighteen to twenty-nine years, who earned between $20,000 and $29,000 annually (2001: 539) — not the demographic portrayed in television's programming. But by far the most striking finding, corroborating Shanahan's previous research, was that the environment was essentially absent from television's programming (about half of one percent). As McComas et al. nicely highlight, in six years of fiction/entertainment programming there wasn't enough environmental programming to outrun one Monday night football game!

More recently, Rowan Howard-Williams (2011) explored environmental television programming in New Zealand. His analysis was very similar to Shanahan and McComas's (1999), but his sample consisted of 140 hours of fiction and non-fiction television programming on four "free-to-air" channels (TV One, TV2, TV3 and Maori TV) in New Zealand. Based on this, Howard-Williams concluded: "Nature was completely absent as a theme from 47 of the 77 non-news programmes… and was a primary theme in just seven (under 10 percent)" (34).

Perhaps not surprisingly, therefore, content analyses of television's prime-time fiction programming show that environmental content is largely absent.

Environmental Cultivation: Effects

Many studies have shown that there is a relationship between mediated communication and how consumers of that media understand environmental issues (Shanahan and McComas provide an excellent overview in *Nature Stories*). In an early example of environmental communication effects research, Kenneth Novic and Peter Sandman (1974) found that "heavy users of mass media [including television] were less informed, viewed environmental problems less seriously, and preferred less 'personal' solutions to environmental problems" (quoted in Shanahan and McComas 1999: 36).

Christine Ader (1995) made use of agenda-setting theory (outlined in the Television chapter) to explore the relationship between coverage of environmental issues in the *New York Times* from 1970 to 1990 and how the public prioritized environmental issues. By calculating the length of an environmental story, defined as a story that explores "humanity's unintentional disruption of the ecological system such as disruption of wastes, air quality and water quality" (303), and the story's prominence, based on where and how the story appeared in the newspaper, Ader ranked the environmental story's importance. This ranking was then compared to the public's answers to the Gallup poll question "What do you think is the most important problem facing the nation today?"

Through this comparison, Ader found a relationship between the prominence of certain environmental issues in the *New York Times* and the importance the public gave to those same environmental issues. In addition, by also looking at "on the ground" — and in the water and air — measures of pollution, Ader was further able to conclude: "The public needs the media to tell them how important an issue the environment is. Individuals do not learn this from real-world cues" (1995: 310). Ader's results supported the agenda-setting hypothesis that the media's agenda influences the public's agenda.

While studies such as Ader's give us valuable insight into the short-term implications for how mediated communication affects our understanding of the environment, cultivation researchers have tried to tap into the implications for our long-term exposure. As Shanahan and McComas highlight, "the term *cultivation* is used to indicate that the process is conceived as a cumulative one; messages and their background contexts are seen as having gradual impacts on audiences repeatedly exposed to them" (1999: 117). Environmental cultivation researchers have sought to explore these cumulative long-term or meta-effects of television's fictional stories as they relate to our understanding of our relationship with the environment.

James Shanahan's first environmental cultivation study took place in the late 1980s and early 1990s. He used college students as research participants and collected data four times between 1988 and 1992. As with all cultivation research, there were measures that tapped into perceptions of, and feelings about, the "real world" as well as measures of television use. In this initial environmental cultivation study, the researchers divided perceptions about the environment into four categories and asked for participants' responses to various statements.

The four categories were: 1) optimism about the environment (measured as level of agreement with statements such as "We shouldn't be too concerned about things like acid rain and the ozone layer, because they will take care of themselves in time"); 2) importance of the environment relative to other world issues (measured as level of agreement with statements such as, "The good things that we get from modern technology are more important than the bad things like pollution that may result"); 3) personal impact/agency in environmental issues (measured as level of agreement with statements such as, "It doesn't matter what I do, the environmental problem is too big for any one person to have any impact"); and 4) attitudes about specific environmental issues (measured as level of agreement with statements such as, "Companies should stop using plastic for food packaging, even if it costs consumers more at the grocery store").

Across the four rounds of data collection, Shanahan found that compared to lighter television viewers, heavier television viewers were less likely to rate the environment as an issue of concern (relative to other issues) and more likely to feel that the future of the environment is secure. Heavier television viewers were also less likely than lighter television viewers to think that, as individuals, they could have an impact on environmental issues (Shanahan and McComas 1999).

In a subsequent study, Shanahan and his colleagues made use of the results from the National Opinion Research Council's (NORC) General Social Survey (GSS). The GSS conducts basic scientific research on the structure and development of American society with a data-collection program designed to monitor societal change within the United States. Each year the GSS collects data from personal interviews with several thousand research participants, and each year one of the things those participants are asked about is their daily television viewing. In 1993 and 1994, there was also a special "module" of questions related to the environment. These two elements, television and the environment, allowed the researchers to undertake an environmental cultivation analysis with a broader research population than had previously been possible. The findings from the research were complex and not always in keeping with the results that the researchers were expecting, but one finding was predictable: heavier viewers of television were less willing to make environmental sacrifices. This corroborated the previous findings that heavier viewers of television are less concerned about the environment (and thus less motivated to do anything about it).

The 2000 GSS once again had a module on the environment, and this time James Shanahan and I undertook an environmental cultivation study in which we compared concepts such as "environmental activism" (joining an environmental organization, signing an environmental petition) and "environmental affirmation" (responses to statements such as "There are more important things to do in life than protect the environment") for heavy and light television viewers. We found that heavier television viewers scored lower than lighter television viewers on both of these measures.

In the early 2000s, as part of my dissertation research, I distributed 2,000 surveys, 1,000 to a random sample of adults in the U.S. and 1,000 to a random sample of members of a U.S.-based environmental organization. The survey tapped into environmental attitudes with the New Environmental Paradigm scale developed by Riley Dunlap, Kent VanLiere, Angela Mertig and Robert Jones (2000). The scale measures what the developers call an "ecological worldview" using the following fifteen statements, to which research participants can answer from 5 for strongly agree to 1 for strongly disagree. (Note that, as with most such scales, the statements go in different "directions," but the results are tabulated such that a higher score on the scale would mean a "stronger" ecological worldview.)

1. We are approaching the limit of the number of people the earth can support.
2. Humans have the right to modify the natural environment to suit their needs.
3. When humans interfere with nature it often produces disastrous consequences.
4. Human ingenuity will insure that we do NOT make the earth unlivable.
5. Humans are severely abusing the environment.
6. The earth has plenty of natural resources if we just learn how to develop them.
7. Plants and animals have as much right as humans to exist.
8. The balance of nature is strong enough to cope with the impacts of modern industrial nations.
9. Despite our special abilities, humans are still subject to the laws of nature.
10. The so-called "ecological crisis" facing humankind has been greatly exaggerated.
11. The earth is like a spaceship with very limited room and resources.
12. Humans were meant to rule over the rest of nature.
13. The balance of nature is very delicate and easily upset.
14. Humans will eventually learn enough about how nature works to be able to control it.

15. If things continue on their present course, we will soon experience a major ecological catastrophe.

I also asked about television viewing in two ways. First, I asked about hour-per-day estimates for both weekday and weekend viewing. Second, I made use of a six-statement scale developed by L. J. Shrum and colleagues, with statements such as "I watch less television than most people I know" and "One of the first things I do in the evening is turn on the television." (The full scale is in the Materialism chapter.) I also asked various demographic details so that I would be able to control for sex, age, income, education, area of residence and first language. As with the cultivation studies highlighted above, my research found that those who watched greater quantities of television scored lower on the New Environmental Paradigm scale (Good 2007). Or, as I offer in the discussion section of the article, when people are heavy viewers of television, they are less likely to see the environment's limits to growth, the fragility of nature's balance and the possibility of environmental crisis. Instead, heavier viewers are more likely to feel that humans are isolated from the rest of the environment and apply anthropomorphic (human-centred) thinking to the non-human world.

Heavier television viewers are also more likely to be fearful about the personal and familial consequences of environmental degradation. For example, Lance Holbert, Nojin Kwak and Dhavan Shah (2003) found that heavier television viewing was related to a higher level of agreement with the statement: "I worry a lot about the effects of environmental pollution on my family's health." More recently Michael Dahlstrom and Dietram Scheufele (2010) found that agreement with the statements "I worry a lot about the effects of environmental pollution on my family's health" and "I'm very concerned about global warming" was higher for those who watch more television – and viewing a larger number of television channels was related to even greater environmental fearfulness.

But Wait a Minute!

When students first encounter these environmental cultivation research findings, they often resist the notion that heavy television viewing can affect environmental attitudes. One of the most common forms of resistance runs along the lines of this question: "But what about my aunt Gloria? She has watched television non-stop her entire life and she's the world's most environmentally aware and concerned person!" Part of the answer to this question is the fact that there are always exceptions (or, in the language of statistics, there are always outliers). In addition recall that cultivation theory is not proposing that heavy viewers are destined to dump toxic waste in the nearest waterway (or, in the language of statistics, effects sizes are small but present and consistent). That said, consistent and small effects, above and beyond all of the other inputs in our lives, and compounded by all of the other people in our lives whose understandings of the world have been affected by television, should be taken seriously.

Another common reaction to the results of environmental cultivation research is a sense of personal confrontation. Students will often say something along the lines of, "Well, good thing I take communication courses so television won't affect me," to which I reply something along the lines of, "Critical thinking can be helpful, but one of the things that cultivation theory highlights is that television's effects are not conscious effects."

As I pointed out in the Television chapter, cultivation research, especially several studies by L.J. Shrum, shows that when people are given the opportunity to "rule out" television as their source of information (e.g., if television viewing questions are asked in a survey *before* questions about the environment), then the cultivation effect often disappears. Shrum's research also shows that the cultivation effect is weaker when people answer questions at their leisure on a written survey than when they answer questions as part of a phone survey — since in the latter case they feel rushed and do not have an opportunity to discount television's effects (2007). In these examples, therefore, our thoughtfulness *is* related to cultivation theory. But if this leads you to think that you'll just "outsmart" television's effects, it doesn't work that way. In fact, recall from the Materialism chapter that while thoughtfulness can erode or even eliminate the cultivation effect in some situations, the research also indicates that it is those who are most caring and thoughtful about the environment who may wind up being the ones *most affected by heavy television viewing.*

In the Television chapter, I introduced "mainstreaming" — a cultivation theory concept that proposes that television encourages viewers to have a common understanding of the world. As such, those who watch only a little television will have disparate views about an issue like the environment (some care a lot, some not at all). When television viewing increases, however, *everyone's attitudes, values and beliefs become more like television's attitudes, values and beliefs.* In other words, heavy television viewers' attitudes, values and beliefs are affected by a similar "gravitational pull" (i.e., television) and "mainstreamed." I explained that women who are light viewers of television tend to think that the world is a meaner and scarier place than men who are light viewers, because women are statistically more likely to be victims of assault. When men and women are *heavy* viewers of television, however, their sense of the world as a mean and scary place becomes more similar, with men's attitudes about safety mirroring women's. This is because these men are learning from television, not their lived reality.

In my research, I have found a similar mainstreaming phenomenon in environmental cultivation. When environmentalists (people who are members of an environmental organization and who self-identify as "environmentalists") are compared to average citizens and both groups are light viewers of television, their average scores on the New Environmental Paradigm scale are quite different (with environmentalists exhibiting higher rates of concern for the environment than average citizens). However, this difference lessens when the members of the two groups watch greater quantities of television. The heavy viewing environmentalists still score higher on the NEP scale than heavy viewing average citizens, but the environmentalists' scores are more similar to the scores of the average citizen. Television has eroded, if you will, the environmentalists' "ecological worldview" — and, presumably, this is not something that environmentalists consciously desire.

Okay, so what about the thoughtful part? What if an especially *thoughtful* environmentalist watches a lot of television? If you recall from the Materialism chapter, this is a question that Shrum asked about materialism and thoughtfulness. His research found that heavy television viewers who scored higher on a need for cognition (NFC) scale were *more* affected by television's materialism than heavy television viewers who scored lower on the NFC scale (Shrum et al. 2005). I have explored the same phenomenon in the context of environmentalism. In order to do this, I made use of the following five-statement scale (first developed by Epstein, Pacini, Dene-Raj and Heier and then modified to a short-form by Petty and Cacioppo):

1. I don't like to have to do a lot of thinking.
2. I try to avoid situations that require thinking in depth about something.
3. I prefer to do something that challenges my thinking than something that requires little thought.
4. I prefer complex to simple problems.
5. Thinking hard and for a long time about something gives me little satisfaction.

The results indicated that those environmentalists who scored higher on the NFC scale had a *stronger* cultivation effect than those environmentalists who scored lower on the NFC scale. I think it's important to explore how to make sense of this finding.

In his research on the cognitive processing of the cultivation effect, L.J. Shrum highlights that there are two fundamental types of information processing we undertake when asked a question: set size estimates and attitudes/values/beliefs. Set size estimates have to do with numbers, facts and figures. When it comes to the environment, set size estimates would be questions about what percent of the world's fish stocks are under immediate threat or how many acres of first growth forest remain on the planet today.

Unless you work in these areas or are particularly interested in environmental statistics, the numbers won't be top of mind. You therefore come up with an estimate by piecing together what you know — from any of a number of sources — about fish or forests. Shrum proposes that television's effects for these set size estimates do not happen *while* viewing but rather in the moment you are forming an estimate. Television has its effect, therefore, *after* viewing. As such, high need for cognition *decreases* the cultivation effect *for* set size estimates. Why? Because when television's information is thoughtfully considered as part of a mix of other sources of information, it is fairly easy to dismiss television's information if it doesn't mesh with the facts and figures obtained via personal experience, friends, family, the newspaper, church, etc. The situation is very different, however, for the effects of television on our attitudes, values and beliefs.

Shrum proposes, and we can intuitively see, that attitudes/values/beliefs are fundamentally different from set size estimates. Our attitudes/values/beliefs are always carried with us and while they change over time, we do not usually form them "on the spot." Take, for example, the NEP scale. If someone is asked to respond to statements such as "Humans were meant to rule over the rest of nature" or "The balance of nature is very delicate and easily upset," our responses don't get compiled in a piece-by-piece logical analysis, but instead we respond to these statements more emotively and at a gut-level.

In this case Shrum proposes that television's effects happen *at the time of viewing*. Shrum also proposes, and my research supports, that for attitudes/values/beliefs, those who score higher on the NFC scale will have a *stronger* cultivation effect. Why wouldn't a thoughtful person use all that "thoughtful energy" to critique and dismiss what is being watched? Because what fun is it to watch television if you're constantly critiquing and dismissing? As I pointed out earlier, if you watch a lot of television it is because, at some level, you enjoy it. And if you're enjoying it, then you're using all that thoughtful energy to *really engage* what is being watched.

The environmental cultivation research therefore makes two things clear. First, watching a lot of television makes us less "environmentally friendly." Second, watching a lot of television is especially problematic for those who care about the environment and like to think. (Thoughtful environmentalists stand to lose the most by watching a lot of television.) Whether we like it or not, therefore, the cultivation research is telling us that there is a relationship between watching television and our attitudes/values/beliefs about the environment. A good question, therefore, is *why*? Why would the amount of television we view be related to how we understand the natural environment?

Why Are Television and Views about the Environment Related?

When I ask students why watching a lot of television makes us less "environmentally friendly," the first answer is usually that those who watch a whole lot of television don't have time to do much else — like be in the outdoors, in the environment. True enough. When we look at the average daily viewing times of just over five hours, we quickly see that when we factor in sleeping and going to school and/or work, there's not much time for anything else. This "time displacement" was immediately obvious when television first arrived in our homes. People deserted all kinds of community activities and stayed home to watch television. Robert Putnam notes in *Bowling Alone: The Collapse and Revival of American Community*:

> This massive change in the way Americans spend our days and nights occurred precisely during the years of generational civic disengagement. How is television viewing related to civic engagement? In a correlational sense, the answer is simple: More television watching means less of virtually every form of civic participation and social involvement. (2001: 228)

Of course in our multi-tasking twenty-first century lives we often do other tasks while watching television — we can text, play games, study — but time spent outdoors does not usually figure into this scenario.

This time displacement aspect of television viewing, therefore, has been offered as one possible explanation for why it is that heavy television viewing is related to "anti-environmental" attitudes, values and beliefs. These kinds of concerns have set in motion calls for increasing the time we spend in the outdoors. Perhaps best known in this vein is Richard Louv's work, including his extremely well received 2005 book *Last Child in the Woods: Saving Our Children from Nature Deficit Disorder*, in which he explores the implications for young people spending so little time in the outdoors. What Louv is able to show, by drawing on many studies, is that young people's social, psychological and spiritual well-being is adversely affected by the absence of play in nature. Louv also highlights the abundant research illustrating the wonderful outcomes for children spending time with nature. Louv's exploration of the importance of time spent in the outdoors encouraged him to launch a non-profit organization, The Children and Nature Network, which encourages kids, and parents, to get outside.

There are other possible explanations, however, for how television affects our relationship with the natural environment. For example, recall the cultivation theory concept of "symbolic annihilation" from the Materialism chapter. What symbolic annihilation proposes is that the quantity of television viewing relates to our understanding of the real world not only because of the stories that television tells us, but also because of the stories that television *doesn't* tell us. In other words, perhaps heavy viewers of television have less concern about the environment because, as we know from the television content analyses I highlighted above, there aren't many stories about the environment on television (especially not in prime-time fiction programming).

However, I have never been completely satisfied with either of these explanations for the relationship between heavy television viewing and "anti-environmentalism." The time displacement argument seems unhelpful because while of course we *could* be doing other things, the reality is that we *are* watching a lot of television. Even if light viewers of television are spending more time outdoors than heavy viewers of television, there remains the question of the consequences of the content that the heavy viewers are consuming. The symbolic annihilation argument similarly seems unhelpful because, while it's true that we don't get a lot of environmental content when we watch television, we do get a lot of *other* content. What role does the content we *do* get from television play in shaping our environmental attitudes?

My lack of satisfaction with time displacement and symbolic annihilation, therefore, set in motion research to collect data to test a third possible explanation for the environmental cultivation phenomenon. My research began in the place that all cultivation research begins, with the question: what is television's story? In this case, rather than asking about how television tells stories about violence, race, gender or the environment, I wanted to explore a story much more fundamental and universal to television. I wanted to explore how television tells stories about materialism. Of course there is research, some of which I highlighted in the Materialism chapter, that has measured specific aspects of television's materialistic messages — and the rate at which we receive them.

There is Neil Postman's calculation that given the rate at which we watch television, we will be exposed to over a million advertisements in our lifetime (1986: 126), and there is Rosellina Ferraro and Rosemary Avery's research that looks at how frequently brands/products appear on, and are referred to in, non-advertising programming (over fifty-eight times in a 30-minute segment for some programming). But in some ways these calculations and analyses do a disservice to television's relationship with materialism.

I believe that the most important aspect of television's relationship with materialism isn't the explicit references to what we should consume — as found in advertisements, brand references, product placement, product integration — but rather what is important about television is that *everything* that television does is about selling us materialism. Television's "meta-story" is materialism. Sure, there are nuances in the kinds of stories that are told — sit-coms, dramas, reality shows, soaps, cartoons — but don't be fooled. Television exists to sell audiences to advertisers such that advertisers sell us not just their stuff but an entire system of logic in which our endless coveting and acquisition of stuff makes sense. Indeed, materialism has become so logical it is common sense. The assumption that we always want more is logical to us, not surprising.

This selling aspect of television is at the heart of the third component of cultivation theory: the cultural context — created by what George Gerbner called the "market-driven mythology" — out of which television is created. Television's cultural context is about powerful business interests, the desire to maintain the status quo and, television's ultimate role, the endless promotion of materialism in the name of lubricating the market economy.

How does this help us to answer the question at the beginning of this section: *Why* are television and our attitudes about the environment related? Because television's meta-story of materialism is the antithesis of environmentalism. As Shanahan and McComas offer in *Nature Stories*, "television programs are designed specifically to serve as frames for advertisements. Advertisements are designed to promote the consumption of products... these goals are fundamentally at odds with the goals of the environmental movement" (1999: 109). Why are the consumption of products and the environment fundamentally at odds? Because materialism endlessly demands a fixation upon the *end results* of the production process. Materialists believe that what they own, wear, drive, etc. is intimately connected to *who they are* and thus they crave products to fulfill who they are; the materials and processes that went into creating the product are but a footnote (if that). Thus, the strain put on the planet's ecosystems to provide an endless stream of cheap consumer goods is not part of the equation. The only consideration is an unrelenting desire to have the latest, greatest, flashiest, most fashionable item.

Based on this, it is not surprising that, as Tim Kasser points out in *The High Price of Materialism*, "substantial evidence shows that choices arising from a materialistic value orientation are often unconcerned with, or actively hostile toward, nature" (2002: 92). E.F. Schumacher succinctly offers in *Small Is Beautiful: Economics as if People Mattered*:

> An attitude of life which seeks fulfillment in the single-minded pursuit of wealth — in short materialism — does not fit into this world, because it contains within itself no limiting principle, while the environment in which it is placed is strictly limited. (1975: 30)

Like others who came before me, therefore, I could intuitively see that materialism and environmentalism were fundamentally at odds. But nobody had ever undertaken an analysis to explore whether materialism could help us understand environmental cultivation. Could materialism provide an answer to the question of why television and our understanding of the environment are related? I undertook research to find out.

As I mentioned above, part of my research involved asking a random sample of Americans and a random sample of members of an environmental organization about their television viewing and their environmental attitudes. I found that there was a relationship between television viewing and scores on the New Environmental Paradigm (NEP) scale such that heavier television viewers tended to score lower on the NEP scale. I also asked the participants about their materialism, and my findings indicate that, on average, the more television a person watched, the more materialistic the person was. Then I tested whether I could show that the materialism explains, or in statistical terms "mediates," the relationship between television viewing and attitudes about the environment.

The statistical analysis involved three steps: first, measuring whether there was a relationship between television viewing and attitudes about the environment (yes); second, measuring whether there was a relationship between television viewing and materialism (yes); third, measuring whether the relationship between television viewing and attitudes about the environment continued to exist when materialism was controlled for (no). According to the statisticians Reuben Baron and David Kenny, when a relationship between variables (i.e., between television and the NEP scale) is statistically significant initially but loses that statistical significance when another variable is controlled for (i.e., materialism), then this controlled-for variable can be understood to "fully mediate" the original relationship (1986). Therefore materialism mediated the relationship between television and environmentalism.

While researchers have therefore tended to see the television and environmental attitudes relationship like this:

Television viewing ⟶ Environmental attitudes

My research proposes that what explains the above relationship is materialism:

Television viewing ⟶ Materialism ⟶ Environmental attitudes

More precisely, previous research had indicated that as television viewing increases, attitudes about the environment become less positive (in this case as indicated by decreasing scores on the NEP scale). My research found that as television viewing increases, levels of materialism also increase (in this case as indicated by scores on the Richins and Dawson materialism scale) and attitudes about the environment become less positive.

If I have been successful in achieving my goals, therefore, I have convinced you of two things: first, we are adversely affected by television; second, television's stories that celebrate materialism are hugely problematic for us and for the planet. Here is one more thing I would like to convince you about: while it is essential to fight against television's role in our planetary crisis, the only way we can win is if we also wage a battle against the story of endless growth.

Blame Television, Materialism and Endless Growth

Context, it is often said, is everything. Under other planetary circumstances, I might be writing about why television's meta-story lauding materialism is unfortunate for all the heavy viewers who are insecure and unhappy (among other ailments) because of their materialism — and the analysis would end there. I could explore how television's constant bombardment encouraging us to understand ourselves as that which we own is playing a role in such things as record levels of personal debt and depression. (A 2011 World Health Organization study found Americans to have the second highest depression rates in the world.) Under other planetary circumstances (for example, if the Earth's ecosystems could instantaneously regenerate to their original form whenever something was removed), the analysis could begin and end with the personal consequences of materialism.

Alas, this is not the planet on which we live. In the twenty-first century we must understand the Earth's resources as finite (and if you're thinking that resources such as wood and fish are "renewable," the state of the world's forests and fisheries should make you reconsider). For much of human history the finiteness of the planet wasn't really a great concern because we lived quite harmoniously on the planet without even remotely testing its limits. But that harmony has recently disappeared. As Peter Victor points out in his article in the journal *Nature*, "the sheer weight of materials, including fuel, that feed the world's economies has increased 800% in the twentieth century" (2010: 370). You might be thinking: "But the world's population has also grown by leaps and bounds in that time." The reality, however, is that while the Earth's population has grown — from approximately 1.6 billion to 7 billion in the twenty-first century — the resource demands have been made by relatively few people; these few people have not only demanded resources, they have demanded endless and ever-growing quantities of resources.

Economist Juliet Schor highlights this phenomenon by looking at how the "personal consumption" portion of the gross domestic product (GDP) has increased for those in the United States in the past few decades. In 1969, U.S. personal consumption was 61.5 percent of GDP (investment, government spending and exports making up the bulk of the rest of the GDP); by 1989, U.S. personal consumption was up to 65.6 percent, and by 2007, personal consumption was over 70 percent of GDP for the first time in history. Therefore, in 2007, the average American's annual personal expenditure (on consumer goods and services) was $32,144. Schor states: "It's an extraordinary figure especially when compared with a global average income of only $8,500, or the fact that more than half the people in the world earn less than $1,000 annually" (2010: 25).

Inequitable distribution is not, however, part of the calculation of the GDP. The planet's very real limitations aren't part of the formula either. In fact, as I highlighted in the Materialism chapter, the GDP has only one criterion for economic health: growth. Human well-being and planetary well-being be damned. That's the thing about television's stories of materialism, they exist in an economic system that values one thing and one thing only: growth. Not growth and well-being, or growth and more equitable distribution, just endless growth at any cost. After all, as Ronald Wright points out, "the great promise of modernity was progress without limit and without end" (2004: 6). So television's endless stories and endless demands for endless growth (or "progress") exist not on some idealized planet but on this very real planet. These endless stories of growth at any cost have become so ingrained in our worldview, so much a part of our "it's just common sense that the economy must grow" thinking, that we are blinded. We are unable to see the illogic — an illogic that is so clear to us in other contexts. Consider, for example, the Ponzi scheme.

You may recall that in the midst of the 2008 economic crisis, Bernie Madoff made headlines because of his massive $50 billion Ponzi scheme (a "borrowing from Peter to pay Paul" financial scam usually done by paying existing investors with incoming investors' money). What I was struck by as I began to hear the details of Madoff's crime was just how incredibly reviled he had become. Indeed, Madoff is understood as a villain extraordinaire. For example, he is likened to a serial killer in the 2009 *New York Times* article, "The Talented Mr. Madoff," by Julie Cresswell and Landon Thomas. Why is Madoff so evil? Not because he made people money (and he did make money — that's why his Ponzi scheme lasted decades and brought in so many new investors) — but *because he made money in a fundamentally unsustainable way.* This is what was illegal about what Madoff was doing — the unsustainability. Indeed, according to a prepared statement by the General Counsel for the U.S. Federal Trade Commission on "Pyramid Schemes" (1998), the reason these scams are illegal is "because *they inevitably must fall apart.* No program can recruit new members forever. Every pyramid or Ponzi scheme collapses because *it cannot expand beyond the size of the earth's population*" (italics added). Hmmm... The more I read about Madoff and Ponzi schemes, the more it seemed like I was reading about our very own endless-growth economic model that cannot expand beyond the size of the Earth's resources...

David Barash comes to a similar conclusion in his *Chronicle of Higher Education* article entitled "We Are All Madoffs: Our Relationship to the Natural World Is a Ponzi Scheme":

Make no mistake: Our current relationship to the world ecosystem is nothing less than a pyramid scheme, of a magnitude that dwarfs anything ever contemplated by Charles Ponzi, who, before Madoff, was the best-known practitioner of that dark art. Modern civilization's exploitation of the natural environment is not unlike the way Madoff exploited his investors, predicated on the illusion that it will always be possible to make future payments owing to yet more exploitation down the road. (2009)

Exactly. On a small scale it's so clear to us. Of course what Madoff was doing was wrong. People who were, in Cresswell and Thomas's words, "hungry for the predictable and handsome returns he booked year after year" trusted him, and he betrayed their trust because sooner or later it was going to become clear that those returns couldn't be sustained. It is less clear to us, on a large scale, that an economy built on endless growth similarly cannot be sustained. Tim Kasser concludes in *The High Price of Materialism*: "The ultimate problem implied by these [materialism] studies is that if we continue to be driven by selfishness and materialism, ecological disaster awaits us" (2002: 93), and, sadly, the evidence compels all of us to agree. So, while it is essential that we hold television, our society's storyteller, accountable for the endless, relentless tales of materialism it spews forth, we must not stop there. The mighty triumvirate of television, materialism and an economic model built entirely on endless growth must be challenged. It is to this challenge that I next turn my attention.

5.   WHAT TO DO?

We are our stories. What we understand about ourselves and our world is based on stories. Stories, therefore, have a deep, fundamental power, and no storyteller has had a more significant role in our lives since the middle of the twentieth century than television. Television's stories may, at the surface, seem varied. There are stories about violence and crime, to be sure, but there are also lighthearted stories about funny families and lovelorn couples; there are even "educational" stories about animals, cooking and gardening. But television's fundamental role is to bring audiences to advertisers; its meta-story, the story on which all of these other stories have been built, is materialism. Materialism is the story that props up and reinforces our endless-growth economic system. If humanity is going to survive, the story of materialism and the endless growth economic system that it propels will have to end.

Turn Off Your Television!

Jerry Mander describes television as "a perfect instrument for the centralized control of information and consciousness" (1991: 3), and because so few people create television's stories while so many absorb its messages, Mander comments that television "freewayizes, suburbanizes and commoditizes human beings, who are then easier to control" (1978: 349). As one of only a handful of outspoken critics of television and perhaps the most outspoken of the bunch, there is much that is compelling and satisfying about Mander's unabashedly biting words.

If we want to diminish television's power as our society's storyteller, then one way is to turn it off and listen to other stories. This is at least part of what Mander was getting at when he wrote a book advocating the elimination of the medium. In 1978, television's elimination was a rather audacious suggestion and seemed so farfetched as to be absurd. Decades later, however, he is in good company. The American Academy of Pediatrics, the Canadian Pediatric Society and the American Medical Association have come out with clear statements about why we should be wary of television (especially for the very young) to the point of avoiding it altogether. It is incredibly significant that these hugely influential medical associations, organizations that are not known for making rash judgments, have cautioned against television. Given the wealth of television research that encourages us to be leery of the medium's effects, caution is not farfetched in the least; indeed it is extremely prudent.

Other organizations' concerns about television have led to similar encouragement to avoid the medium. The activist organization Adbusters Media Foundation launched a "TV Turnoff Week" as part of what it calls "culture jamming" activities. Their 30-second ad ran with a black screen and the following voiceover:

> Do you spend more time watching nature programs than experiencing the real thing? More time laughing at TV jokes than joking around yourself? More time watching TV sex than making love yourself? Then there's something you're missing. Television isn't real; but the addiction is. (Adbusters 1993)

In 1994, the non-profit organization TV-Free America was founded, and it now runs its own National TV Turnoff Week. Its website lists fifty-two alternative non-TV activities one can undertake, such as volunteer in a school, learn to play a musical instrument, attend community concerts, organize a community clean-up, put together a puzzle, visit the library, go skating, visit the zoo. More recently, this message has expanded to include turning off other types of screens as well. The Campaign for a Commercial-Free Childhood promotes Screen-Free Week, asking people to "Turn off screens... and turn on the world around you." Adbusters' TV Turnoff Week has morphed into Digital Detox Week.

If there's one thing that cultivation researchers have taught us over the years it's that we should never underestimate the power of television's storytelling. Television's version of the world not only helps form the stories we tell about ourselves and our world, but television's version of the world also helps form stories for our family, friends, co-workers, politicians, business leaders — everyone. If we are going to change what we value in our lives, such that we are less materialistic and tread less heavily on the planet, we are going to have to fundamentally challenge those stories. Television's stories need to be refused.

Does that sound like crazy talk to you? If you're anything like the thousands of students I've interacted with over the past twenty years, then the idea of doing away with television's stories (not just the stories as they appear on the traditional television but also as they appear on other types of screens) seems comparable to asking someone to go without air. How could we possibly live without television's stories? Besides, students are quick to wonder (it can almost take the form of a plea), can't we just use television to tell *better* stories? In what follows I answer these questions in four steps. First, I explore why it's not so crazy to envision a television-free world. Second, I offer why "better" stories on television aren't better. Third, I discuss the essential role of sharing our stories. Finally, I look at the way in which media literacy can help move us to a post-television, post-materialistic and environmentally robust world.

Television Is Incredibly New

My exploration into why we should believe we can kick the television habit begins with a baby. When a baby enters the world, everything is measured in inches — what she can grab, how far she can see, how close she wants her mother's milk to be at all times. There is nothing more to the world than those few inches. Our understanding of how our lives fit into humanity's history is often baby-like. Without a long view of history, we can miss that the past few hundred years have been anomalies. Mediated communication is brand new, television newer still. The endless coveting of material goods is similarly new and something that mediated communication had to help us learn. Our current relationship with the environment is also new and markedly sets us apart from the tens of thousands of generations that came before us.

Like a baby, therefore, it is easy for us to only experience and see our history in inches. It is easy for us to believe that "what is" has always been and is all that will ever be, but the reality is that we're not in this place because the long experiment of human history pointed us here. In fact, if anything, the past few hundred years have been an aberration. This is good news for two reasons: first, because our current way of doing things is new and from the perspective of human history, not so ingrained; second, because if there is anything that the past few hundred years shows us, it is that we are very capable of change. It is not, however, only large-scale social changes over the past few hundred years that illustrate our capacity for new ways of doing things — we have more recent examples as well.

Consider cigarettes. Like our relationship with television, our relationship with smoking began recently. In *Smoke: A Global History of Smoking*, Sander Gilman and Xun Zhou offer:

> Advertising imagery became the place where twentieth-century fantasies about smoking to provide status, desirability or pleasure were defined. With every purchase, consumers were offered an intense yet vague, magical yet repeatable promise: they too could be like the smokers in the advertisements. (2004: 21)

And the advertisements *worked*. We were won over. According to Gallup, 45 percent of Americans were smoking in 1954 and anyone who lived in the 1950s, '60s and '70s knows from first-hand experience that society was set up to accommodate those who smoked — and to encourage everyone else to join in (Saad 2008). My parents were both smokers, and they would often highlight the fact that when they were growing up in the 1950s, doctors promoted smoking on television. When I was growing up in the 1970s, cigarettes were smoked everywhere: offices, restaurants, buses, planes and even hospitals. But something happened to smoking in North America. What was unfathomable at the time became the law as smoking bans in offices, restaurants, buses, planes, hospitals and even bars have become increasingly prevalent.

The changes were not without contention, and there are those who continue to smoke and fight for the rights of smokers, but here in North America what was once a deeply entrenched reality has dramatically changed and changed quickly. The policies and laws that have gone into changing the social reality of smoking are numerous, but a big part of the facilitation of change had to do with shifts in the stories that were told, or in this case, no longer told, about smoking. For example, in April 1970, the U.S. Congress passed the Public Health Cigarette Smoking Act, which banned cigarette advertising on television and radio. The ban commenced in January 1971 and continues to this day. Other policies have since been put in place to facilitate society's changing relationship with smoking — enforced minimum smoking ages, graphic warning labels on cigarette packages, anti-smoking messaging, etc. — and these efforts also continue, with varying levels of success, to this day.

Smoking was an incredibly widespread, not to mention incredibly addictive, social phenomenon that came under attack. What the somewhat brief history of smoking exemplifies, therefore, is the power of stories to shape our reality and overcome obstacles to change. Gilman and Zhou point out in *Smoke*: "Be it a panacea or bane, icon or commodity, the magic of smoking and the images surrounding it continue to shape and be shaped by our changing perception of the world" (2004: 28). Smoking was, and remains, a story, and the beautiful thing about stories is that they can be changed.

Is television too tangled in our lives to ever be removed? The smoking analogy indicates that even some of the most tangled and entrenched practices can be fundamentally changed. Many people believed that they could never possibly live without smoking. Yet many have quit (not everyone, by any means), and the culture of smoking has changed beyond what we would have once thought possible.

Still, not watching television sounds so hard. What if instead of stopping television viewing we changed television's stories? Can't we just continue to watch but tell *better* stories? The answer is no, and ironically the case for this answer can be found in the few attempts that have been made to do just that.

Failure of Television's "Better Stories"

When faced with the challenges presented by having television as our society's storyteller, people often become excited by the possibility that television could tell different stories. Television could tell better stories. This is certainly the option that many university students find the most exciting. The "different content" approach is especially appealing to people born in the late twentieth and early twenty-first centuries because they so often feel a fundamental impossibility of getting rid of television — even from their bedroom for one night, let alone for a week or more! It makes sense that television looms so large for this generation given that many of them have had a television screen as an active presence in their lives pretty much from the moment they were born. For them, a very appealing logic goes: "If television's stories can affect us and our values such that the planet is adversely affected, then why don't we change the stories and have television tell stories that positively affect us and the planet?"

As I mentioned in the Television chapter, *Sesame Street* is often held up as the example of television programming that can positively influence viewers, in this case very young viewers. In their overview of thirty years of research on the impact of *Sesame Street* (a program they call the most studied in television history), Shalom Fisch, Rosemarie Truglio and Charlotte Cole conclude that the show has been "effective in encouraging (for example) literacy, prosocial behavior [e.g., sharing and cooperation], mathematics skills, race relations, and preschoolers' understanding of death" (1999: 186). The creators of all kinds of children's television programming have tried to model *Sesame Street* and join this category of "valuable television." But forty years later *Sesame Street* remains the shining example of an ongoing collaboration between the creators of television and educational experts.

In honour of the program's fortieth birthday, that collaboration between television and educational experts focused on a two-year "My World Is Greening and Growing" environmental theme. Of course there are also examples of good quality environmental television for adults. The 2006 Discovery television series Planet Earth is one example. Alerting viewers to "prepare to see the Earth as never before," the well-received series offers "revolutionary filmmaking" that "brings the planet to life." And there is reason to believe that such non-fictional television can be beneficial. Some of my research, and that of other researchers (e.g., Holbert, Kwak and Shah), suggests that heavy viewing of non-fictional programming (e.g., nature documentaries) can be related to *greater* concern for the environment. However, I offer two caveats. First, unlike environmental cultivation research, these non-fiction television studies have not explored causation (i.e., perhaps those who are more concerned about the environment are more likely to watch environmental programming — this is my guess). Second, as cultivation researchers point out, what most of the people watch most of the time is fictional programming.

Based on this reality, some organizations have jumped on this notion of television as a potentially pro-environmental force and have tried to work with producers of prime-time fiction programming to increase the quantities of environmental content. For example, the Environmental Media Association

(EMA), founded in 1989, believes that by weaving environmental messages into prime-time television, it "has the power to influence the environmental awareness of millions of people" (EMA website). In an article highlighting the work of the EMA, Kivi Leroux offers examples:

> A carton of recycled copier paper sits on the counter of the *ER* nurses' station. The cast of *Friends* pours milk out of a reusable glass bottle. *Law and Order*'s Detective Briscoe asks his lieutenant to guess what the blue fleece found at the crime scene is made of. "Recycled plastic bottles," she responds. From props in your favorite star's hands to stories about energy conservation and pesticides, environmental products and themes are appearing regularly on America's most popular television programs [because of the work of the EMA]. (Leroux 1999)

Does such in-program prime-time environmental messaging have an impact? Apparently Al Gore thinks so as he declared: "No group has had a larger impact [than the EMA] on the thinking Americans bring to the environment, on the way we, as a nation, converse with the problems that beset the environment" (Leroux 1999). Leroux uses historical examples — such as library card requests increasing by 500 percent in 1977 in the U.S. after the Fonz applied for a card in an episode of *Happy Days*, and a 67 percent recognition rate of "designated driver" a year after the new term was introduced in over a 160 prime-time television episodes — to illustrate the success of television's socially positive persuasion.

Another approach the EMA has undertaken to spread environmental awareness has been to laud celebrities who speak out for the environment and display "environmentally friendly" behaviours. Each year the EMA has an environmentally themed award show complete with a green carpet on which celebrities walk after arriving in a hybrid or alternative fuel vehicle. Indeed, some celebrities' environmental activities have become well known to us. Film star Leonardo DiCaprio is often held up as an icon of celebrity environmentalism. He is associated with a wide range of environmental activities, including producing and narrating the 2007 climate change documentary *11th Hour*, and, through the Leonardo DiCaprio Foundation, he supports "efforts to secure a sustainable future for our planet and all its inhabitants" (Leonardo DiCaprio Foundation website). In the television world there are various celebrity environmentalists. Julia Louis-Dreyfus, star of *Seinfeld* and more recently *The New Adventures of Old Christine* and *Veep*, is one such TV enviro-celebrity — referred to as an "eco-warrior" and "devout environmentalist" by the online environmental magazine *Grist* (Little 2003) for her embracing of environmentally thoughtful living.

But here's the problem with celebrating television in any context: it's easy for the drops of "good stuff" to distract us from the ocean of materialism. So, while the research on *Sesame Street* is clear — children (especially from more affluent and well-educated homes) can learn valuable lessons from their viewing — there is a much more fundamental lifelong lesson children are gaining: how to watch, and yes, learn from television's meta-story. And this meta-story is not about love for the environment; it is about materialism. Indeed the case can be made that our celebration of *Sesame Street* led to the phenomenon of parents encouraging television viewing, in the name of "education," at an earlier and earlier age. As a somewhat revolutionary *Journal of Pediatrics* article by Frederick Zimmerman, Dimitria Christakis and Andrew Meltzoff highlights, the viewing of such early "educational" videos, like *Baby Einstein*, is associated with delays in language skills. In fact, the authors conclude that for infants (age eight to sixteen months), each hour per day of viewing baby DVDs/videos was associated with a 16.99-point decrease on the Communicative Development Inventory — the equivalent of about seven fewer vocabulary words for each hour of viewing (2004).

Therefore, while learning useful lessons from *Sesame Street* is certainly possible, our embracing of television's stories is, in general, extremely unwise. This is also true of the efforts of an organization like the EMA. Yes, there can be moments of environmental messaging on television and there can be celebrity environmentalists who pass along important information, but these bits of information exist within a much bigger, and louder, story of consumption and materialism. What decades of cultivation theory's systematic analyses have shown us is that television's prime-time fiction environmental content is very limited. Maybe a glass milk bottle or a carton of recycled paper appears here and there, but in the ocean of television's stories, these are drops. Television is, almost without exception, a commercial medium that tells us stories about our inability to have, or be, "enough" and then sells us stuff that can, at least temporarily, fill the endless want. Even the most well-intentioned message and messenger exist as fodder for the larger meta-story of yearning for more. Of never having enough.

Take, for example, the *Grist* article mentioned above, which lauds the great model of environmentalism television celebrity Julia Louis-Dreyfus offers. The article encourages us to celebrate the effort to which she and her husband went in building their second home — a beautiful place on the ocean complete with a retractable roof and four cars (three of which are energy efficient) in the driveway. Granted, enviro-celebrities seem somewhat self-conscious about the inconsistencies. The couple acknowledges:

> Having a second home is itself a sort of appalling excess. We figured
> if we're going to do it, we better be as responsible as we can....
> Admittedly, some of the tropical hardwoods — the bamboo and

sustainably harvested mahogany and ipe [a rainforest hardwood] — were shipped in from countries like Guatemala and Honduras. Oh, and there are a few items like the Jacuzzi that kick up the energy load. (Little 2003)

They also say: "As Americans we are incredibly fortunate. We are 5 percent of the world's population and consume a third of the total resources. We should all feel guilty relative to the world." But guilt clearly didn't stand in the way of another home, or a few more cars, or a hot tub — guilt added the solar panels on the retractable roof and meant that at least a couple of the cars had Japanese hybrid technology. You'll grant me that the environmental message is somewhat convoluted. The EMA and Al Gore want us to believe that television can be a place for environmentalism, that celebrities can model that environmentalism. But the EMA and Gore are wrong. It's not that television is unable tell other stories. It is technologically possible for television to tell all kinds of stories. But television's entire existence is based on one fundamental goal: selling audiences to advertisers so that advertisers can sell us stuff. Somehow television's stories, even those people who "represent" television's stories — e.g., the actors — must help achieve that goal.

So, where does this leave us? Can we live without television? Yes. Smoking provides us with one example of what radical social change can look like. We only just learned to live *with* television — we can unlearn. Can we just use television to tell better stories? No. What "better stories" like *Sesame Street* teach us is to watch and learn from television. And as we continue to watch and learn from television, its materialistic meta-story takes over. Cultivation theory's findings show us that the drops of television's "better stories" don't affect our learning of the meta-story: that we are what we consume, and we can never consume enough. What do we do in the meantime as we unlearn television and materialism? We share our stories and embrace media literacy.

It's Hard to Talk about Materialism

We have learned our role as materialists well. According to data from the U.S. Census Bureau and the American Apparel & Footwear Association, in 1991, Americans bought an average of thirty-four items of clothing (pants, shirts, sweaters, jackets, underwear, etc.). In 1996, that number rose to forty-one clothing purchases. By 2007, the average American was buying sixty-seven clothing items each year. This means that in 2007 Americans were purchasing a new piece of clothing every 5.4 days (Schor 2010: 29). How could we have such compelling evidence that we are destroying the very foundations on which our lives depend and yet chip away ever faster at those foundations in the name of jeans, dresses and t-shirts? The answer I am proposing in this book is that it comes down to the stories we're told and the reality we've created out of those stories. Our current reality also has to do with the stories we *don't* tell. Consider the following.

When I was a graduate student, I attended a talk by documentary film-maker Michael Moore. Given that Mr. Moore is one of the most visible and celebrated social activists of our time, my friends and I were excited to hear what he had to say and what he would "be like" in person. Of course we were able to anticipate the kinds of things he'd talk about because of his films. In *Roger and Me*, Moore's 1989 documentary, he took on the issue of big corporations — specifically the automotive giant General Motors — squash-ing the little people. In *Bowling for Columbine*, Moore's 2002 documentary, he explored how big corporations, such as military weapons producers, are related to "on-the-ground" violence such as the 1983 shootings in Columbine, Colorado. The sell-out crowd of about 900 gave Moore a standing ovation when he stepped on stage. He was a bright, engaging and invigorating speaker as he encouraged us to think critically about the world and respond to injustice. When he finished his talk he received another standing ovation. Then something surprising happened.

When the applause ended Moore indicated that he would take questions in a rapid-fire format where questions had to be short and his answers would be less than ten words. One question came from someone in the balcony who asked a question along the lines of, "I understand you own a two mil-lion dollar home. How do you make sense of that?" Moore paused and then flippantly said, "It's a three million dollar home." Full stop. An uncomfort-able rustling went around the auditorium. Moore was ready to move on to the next question, but there were a few boos and hisses. "You didn't like my answer?" Moore asked, somewhat agitatedly, and when people made it clear that, indeed, they had not liked his answer, he continued to respond or, more accurately, to spew. Moore fumed aloud why it was always the "rich fu@ks," who asked things like that, and in general Moore was rude, belligerent and utterly defensive. It was clear that he was extremely upset that he had been asked such a question. I was utterly disappointed by his response.

In discussions and debates I had with friends after Moore's talk, and there were many, some people said I shouldn't have been surprised by Moore's demeanour. After all, such obnoxiousness was on display in his films— in fact being obnoxious is pretty much his trademark and a big part of what makes him such an effective social activist filmmaker. Some said that we shouldn't expect Moore's private life to be up for questioning. And so on. But I remained completely unconvinced. I think that Moore's response to that question and my friends' desire to make excuses for him highlight something extremely significant about our inability to make connections with materialism. Michael Moore has become a highly recognizable and successful activist because he asks difficult questions that many people want to hear asked — and answered.

We celebrate that he does this such that hundreds rise to their feet to give him a standing ovation merely for arriving on the stage. Many people love that he doesn't shy away from researching and sharing the hard stuff.

Yet a question about his multi-million dollar home is disregarded by Moore as being irrelevant and clearly only something a "rich fu@k" would ask in order to provoke him. My discomfort was not about the price of Moore's house *per se*. There may be many good reasons for him to have a multi-million dollar home (it's not even a particularly expensive home in many areas of the U.S.), and presumably he has worked hard for his money. I'll even grant that the question was probably asked by someone wanting to rattle Moore (and it worked). What I wished for in Moore's response wasn't an apology for the value of his home or some kind of confession about his wealth. What I was really hoping for was an acknowledgement, an exploration, of the dominant role materialism plays in our lives.

Moore could have talked about materialism's central role in his various films (e.g., how *Roger and Me* highlights that materialism motivates big business moguls like GM's Roger Smith or how *Bowling for Columbine* shows materialism in the lives of the Lockheed Martin executives he talks about). Moore could have also talked about materialism in his own life. How has his access to wealth affected him and his family? Maybe Moore needs an expensive security system now because he is so hated by some for speaking out. Maybe Moore operates a massive philanthropic non-profit organization out of his home and needs the space. Or maybe he, like most of us, couldn't resist the temptation that came with having some money. Maybe he simply wanted to spend his money on comfort and luxury, including a three million dollar McMansion (as the increasingly large homes in which Americans live have been nicknamed).

However, by refusing us, his 900 audience members, stories that thoughtfully explored materialism and instead nastily turning on the questioner (and, by association, many of us who wanted to hear a thoughtful answer), he silenced what I think is one of the most important dialogues we as a society need to have. Moore's response highlighted what we're up against: nobody really wants to talk about our complicated and often conflicted love affair with materialism. Or, perhaps more precisely, nobody really knows *how* to talk about the love affair. But in a world of rampant materialism, in a world where we are being appealed to as consumers constantly and endlessly — while the environmental foundations on which these appeals are made crumbles — we'd better start talking. We'd better start taking every opportunity we can to explore our relationship with the material world and to create new stories about how we would like that relationship to look in the future. No matter how uncomfortable it makes us. We need to practise.

Practising Our Anti-Materialism Stories

About ten years after my disappointing experience with Michael Moore's refusal to thoughtfully answer a question related to materialism, he released a film that, in some ways, thoughtfully answered the question for him. In 2010, Moore released *Capitalism: A Love Story*, which, as with his other films, he wrote and directed. What Moore highlights in the movie is that while capitalism seemed to serve Americans well for a period of time — arguably providing for us both materially and emotionally — its utility is failing. Why? Moore offers the following excerpt from a speech by President Jimmy Carter on July 15, 1979:

> We are at a turning point in our history. Too many of us now tend to worship self-indulgence and consumption. Human identity is no longer defined by what one does but by what one owns. This is not a message of happiness or reassurance, but it is the truth and it is a warning.

Moore goes on to note that with the instatement of President Ronald Reagan, Carter's warning came to fruition. As most wages stagnated, CEO pay and corporate profits increased — so too did household debt, personal bankruptcies, incarcerations, anti-depressant use and health care costs. All of this set the stage for the market meltdown of 2008. The rest of Moore's movie explores stories that we don't tend to hear. These are stories that critique capitalism, for example, a story about companies taking out secret life insurance policies so that if their employee dies, the company reaps the "benefit" of the policy. Moore also offers a series of interviews with religious leaders. One priest states that he believes capitalism is a sin. "For me and for many of us, at this present moment, capitalism is an evil. It's contrary to all that's good. It's contrary to the common good. It's contrary to compassion. It's contrary to all of the major religions." Moore also tells alternative stories, about companies that don't operate in the typical hierarchical and entirely profit-focused way that capitalism encourages.

As the film ends, Barrack Obama has become president, and Moore highlights the hope that his arrival at the White House brought to so many of us who were longing for an alternative way of understanding our lives and the world in which we live. Moore concludes:

> We live in the richest country in the world. We all deserve a decent job, health care, a good education, a home to call our own… and it's a crime that we don't have it. And we never will as long as we have a system that enriches the few at the expense of the many. Capitalism is an evil and you cannot regulate evil. You have to eliminate it and replace it with something that is good for all people.

And, I would add, with something that is good for the planet.

We need to change the greed-rewarding endless-growth model of capitalism. The place to start is by challenging the stories we are told about the way we should live — and by creating our own stories. Who knows if I now asked Michael Moore about his multi-million dollar home if he would give a more thoughtful answer than he did years ago, but his more recent work gives me hope that he would. And if so, I can only be happy about his change of heart, his increase in thoughtfulness and his desire to tell new stories about materialism.

Stories That Link Materialism and the Environment

Television is our storyteller. As we watch hours each day we are told all kinds of stories, but television's most important and omnipresent story is about how we should understand our relationship with the material world in which we live. Therefore, when that material world gets shaken, as it did during the market meltdown — or economic crisis — of 2008, it is interesting to observe how the storytellers respond. If you recall, the endless-growth model of market capitalism hit a bump. It was a devastating time for many people. Perhaps the only silver lining, but a significant silver lining, was that our economic system was, at least in theory, up for analysis. There was a forced opportunity to tell different stories about our relationship with the material world, but the opportunity was almost completely lost.

The only real solution that was offered by those in power was to prop up the deflated economic structure with infusions of money where extreme greed had caused it to sag, and then to carry on. The success of this monetary bailout was measured by the traditional indicators of economic growth — essentially the rate at which the GDP was increasing. There were, however, a few voices (tragically few, but a few) demanding that we change this narrative and understand the 2008 meltdown as a fundamental indicator of capitalism's failure. Naomi Klein, well known Canadian social activist and author, pondered in a speech she gave in honour of *The Progressive* magazine's hundredth anniversary: "What if the bailout actually works, what if the financial sector is saved and the economy returns to the course it was on before the crisis struck? Is that what we want? And what would that world look like?" (2009). She powerfully concluded that while capitalism could weather the meltdown, "the world can't survive another capitalist comeback."

But come back it has. Huge sums of money were pumped into national economies everywhere, and the North American economy sputtered back to life. The storyline, therefore, continued unaltered: economic growth is the only way, the only solution. There were very few stories that approached the subject from a "maybe our economic model is flawed" perspective, and even fewer that linked the economic crisis and the endless-growth greed that had caused it, to the ongoing destruction of the environment.

I did find one example of a story about the economic crisis that involved the environment. It was a radio broadcast at the start of October 2008 (in the heat of the economic crisis). Canadian scholar, environmentalist and activist David Suzuki was guest hosting the Canadian Broadcast Corporation (CBC) radio program *The Current*. In the show he highlighted the way in which what had seemed to be a fairly widespread discourse of environmental concern, especially around climate change, had been fully derailed by the global economic fears. One of his guests, economist Peter Victor, author of *Managing Without Growth: Slower by Design, Not Disaster*, concurred:

> The downside of [our complete focus on the economy] is of course that we may lose sight of some of the bigger issues that are looming pretty large out there… environmental issues seem to have taken a back seat once again. I think that's very unfortunate. (Victor 2008b)

In the same episode of *The Current* Suzuki interviewed environmentalist Bill McKibben who offered:

> This is a wake up moment — this economic crisis that we find ourselves in is a reminder of how dumb it was to just assume that markets would just take care of problems themselves forever and ever and we would never really have to do anything again — except count the receipts. (McKibben 2008)

Suzuki concluded his discussions by saying: "In many ways this economic crisis is really an opportunity to begin to put these things [the environment and the economy] together." Unfortunately, the stories we were told about the 2008 economic crisis seemed to rarely take advantage of that opportunity.

The fact that I happened upon this one example of voices calling for a critical look at the economic crisis, however, made me want to dig a little deeper. Perhaps I had just missed lots of other examples of the economic crisis being explored from an environmental context. In order to investigate, I undertook a content analysis using the Lexis Nexis database (a searchable news content database). I searched television and radio transcripts from the following television and radio sources: ABC News, American Public Media, CBS News, CNBC, CNN, CTV Television, Fox News Network, MSNBC and NPR. First, I searched "economic crisis" in the subject terms for all stories that had run from the start of September 2008 to the end of December 2008, and I found 4,611 stories. When I expanded the search terms to look for stories with "economic crisis" and any word beginning with "enviro" (e.g., environment, environmental, environmentalism, environmentalist, etc.) in the subject terms, the result was eighty-seven stories. This means that of the television and radio news broadcasts and shows contained within this database that highlighted the "economic crisis" in some way, less than 2 percent also addressed the topic of the environment.

The findings are not surprising, but they do highlight one exceedingly important piece in our deconstruction of our current economic model (and its endless promotion of materialism): we must make sure that there are stories being told about the connection between the current endless-growth economic model and its devastating impact on the Earth. Economist Peter Victor is one person who is offering clear, detailed economic and social analyses that make two fundamental points very clear. First, our economic system must cease using growth as its sole criterion for economic health because if we continue to follow the business-as-usual model, disaster awaits us. Second, there are alternatives. Victor proposes:

> There are indeed feasible economic alternatives [to our current endless-growth model] but getting to them will be beyond us unless we change how we think about our economy, society and environment, undertake some close reflection on what is important to ourselves and others, including other species, and develop a readiness to rethink and transform much of what we have come to take for granted. (2008a: 223)

Economist Juliet Schor is another important voice calling for alternative economic stories and visions. In *Plentitude: The New Economics of True Wealth*, she beautifully illustrates how we might go about developing this "readiness to rethink and transform" the endless-growth model. In fact, Schor gives us some ingredients for creating new economic stories. One key ingredient is time. What role does time play in creating economic change? Let me explain with a story. This story is about a manufacturing company from the U.S. that, like many before it, moves its production to another country where labour costs are cheaper and, as such, the manufacturing of the product is cheaper. A few weeks into the first month of operations, part way through the day, the workers begin to pack up to leave. When those in charge start inquiring about what is going on, the workers reply that they have calculated how much of the month they must work in order to earn the money they need. According to their calculations, they had earned enough money for that month and were heading home to do other things (presumably the things that made them feel happier and more fulfilled than what they were doing at work). They would return at the start of the next month.

Part of what has always struck me about this story is how utterly incomprehensible it was to me when I first heard it (you may be having the same response right now). What would "enough money" even look like in our lives? Could you define it for yourself? You can certainly be excused if you find arriving at such a definition difficult if not impossible because if there is one story we have been told by television (corroborated by other mediated communication and often by those around us) it is that it is impossible to have *enough*.

The above story is apocryphal for several reasons, but here's the most important one: workers in other parts of the world are expected to work *more*, not less, than workers in North America. Getting up and walking out because of having "enough" money is not an option. Schor points out in her book *The Overworked American: The Unexpected Decline of Leisure*, "American textile workers, who enjoy paid vacations and official five-day weeks, are rapidly losing out to their counterparts in China [and elsewhere] where daily, weekly, and hourly schedules are far more arduous" (1992: 52). Workers anywhere who aren't ready and willing to do what is expected of them (e.g., endless hours day-in and day-out) will be workers no more. Nonetheless, just the surprising idea that there might be people, somewhere, who could have such a clear notion of "enough," a concept that is so foreign and otherworldly to us, is noteworthy.

In addition to the almost mythical impossibility of defining "enough" in our lives, this story highlights another important aspect of time and materialism. The promise that came as part of our relationship with the "stuff" in our lives — the relationship that we were promised initially as a result of the Industrial Revolution and then with earnest after the Second World War — was leisure. Leisure, one could argue, has been *the* promise of modern society. Technological "progress" was going to bring us all kinds of gleaming gadgets and whirring devices that would save us the labour and time associated with cleaning, transporting, eating, communicating and so on. Certainly in the more affluent parts of the world, we do have more gadgets and devices than ever before (our homes and landfills attest to it), but as Schor highlights, the research shows that while there was a period between the mid-nineteenth to the mid-twentieth century in which average working hours decreased, since then working hours have increased — and average leisure time (vacation, holidays, paid leaves) has decreased.

So with all of this wondrous time-saving technology, why are we working more and "leisure-ing" less? Schor's analysis gives some credence to employer demands for employees to work increasing hours, but she focuses on what she calls the cycle, or culture, of "work and spend." This cycle, or culture, is based on the encouragement that people receive everywhere in our television-saturated (and generally mediated) modern society to "buy more" and "have more," and, although not as explicitly articulated in the messages, to spend more such that we have to work more to pay for it all. This cycle is a kind of treadmill that requires long hours of work just to remain on it.

Maybe we get into this work-and-spend cycle because it makes us happy.

If we like to work a lot and then spend a lot, even if it puts us in debt, then good for us. There are at least two responses to this: first, it doesn't make us happy, and second, we're not the only ones affected by the work-and-spend cycle — the Earth is being destroyed by it.

We know anecdotally that money doesn't buy happiness — at least that's how the expression goes — and research confirms this. Bruno Frey and Alois Stutzer point out in their book *Happiness and Economics* that numerous studies have "identified a striking and curious relationship: per capita income in the United States has risen sharply in recent decades, but the proportion of persons considering themselves to be 'very happy' has fallen over the same period" (2002: 76). This is not to say that the researchers, or their findings, dismiss the value of a certain fundamental level of well-being that comes from access to basic material comforts, but the notion that more is always better is erroneous. As Robert Lane highlights in *The Loss of Happiness in Market Economies*:

> All the evidence shows that although more money to the poor decreases their unhappiness and also increases their satisfaction with their lives... above the poverty line this relationship between level of income and level of subjective well-being is weak or nonexistent. Thus, the rich are no more satisfied with their lives than the merely comfortable, who in turn are only slightly, if at all, more satisfied with their lives than the lower middle classes. (2000: 16)

In *Affluenza: The All-Consuming Epidemic*, John DeGraaf, David Wann and Thomas Naylor find that while traditional economic measures of wealth, such as the GDP, show general growth in recent history, alternative measures that include well-being in their calculation, such as the Genuine Progress Indicator (GPI), show general decline (2001: 8).

So, let's see. We're working more. We have less "free" time. And we're not very happy. Why do we keep at it? Why do we stay in this work-and-spend cycle that makes us overworked and unhappy? Partly because a treadmill is always in motion, and, therefore, it is very difficult to get off. But we also stay on the treadmill because we have been told so few stories about, and given so few pictures of, what the alternative looks like. This is where changing our ideas about time comes in and where Schor's alternative story — something she calls "plentitude" — comes in. "Millions of Americans have lost control over the basic rhythm of their daily lives. They work too much, eat too quickly, socialize too little, drive and sit in traffic for too many hours, don't get enough sleep, and feel harried too much of the time" (2010: 103). Sound familiar?

This may not be everyone's reality, but Schor's research shows that in general we share what she calls "temporal impoverishment" (noting that in North America we work considerably more than all similarly wealthy countries except Japan). We can, however, turn this around and strive for what she calls "time wealth." A big part of achieving greater time wealth is arriving at an understanding of what it would mean to have "enough."

When we strive for "enough," rather than endless acquisition, we will covet and consume fewer material goods. We will spend less money, we won't need to earn as much money, and the footprint we leave on the Earth will be smaller. Does it seem like an unrealistic dream that we could ever arrive at "enough"? Schor has found that increasingly people are making the choice to earn less. For example, in 1996, 19 percent of adults in the U.S. made choices that led to making less money, and by 2004, that number had increased to 48 percent (2010: 107). Schor's research also shows that 83 percent of the people who made these voluntary shifts said they were happier as a result of their new situation, with only 10 percent saying they regretted the change (108).

Greater time wealth therefore sets in motion various desirable changes. With greater time wealth, we can concentrate on activities that we enjoy and that give our lives meaning. We are also able to thoughtfully take on activities for ourselves and our families for which we had previously relied on the market. This movement, known by such names as "self-provisioning," "urban homesteading" and "DIY" (do it yourself), encourages us to do things for ourselves that we currently pay others to do. With greater time we might choose to grow, prepare and preserve more of our food. We could knit, sew and repair more of our clothes. We might learn to do basic repairs on our homes.

Is this just a version of asking people to leave behind "progress" and return to living in caves? (An argument I hate, but one that gets thrown around all the time.) The answer is a resounding "no." Our movement away from the market economy and our embracing of "self-provisioning" need not be an archaic experience. Indeed it can, perhaps should, be a decidedly twenty-first-century experience. Technology need not be shunned just because we will be involved with a greater diversity of activities directly related to our lives. "It's the perfect synthesis," offers Schor. "Technology obviates the arduous and back-breaking labor of the preindustrial. Artisanal labor [such as is involved with DIY projects] avoids the alienation of the modern factory and office" (127).

As we tell new stories about materialism, we will also continue to take advantage of the very twenty-first-century phenomenon of social networking to do things like decrease our consumption through sharing (and then tell our tales of such experiences). The library is probably the best known, and longest standing, example of sharing — why buy a book when you can borrow it? — but new sharing examples appear all the time. Today digital communication technology has facilitated the sharing of car rides (coordinated through companies such as eRideShare and PickupPal), sharing cars (a much more flexible version of car renting by companies such as AutoShare and Zipcar), sharing bikes (sometimes for free — such as in the experimental Yellow Bike and Orange Bike projects — but increasingly for rent with flexible pick-up/drop-off locations). Indeed, we can share and give away whatever we want through programs such as neighborgoods.com and freecycle.com.

In addition to less consumption, the research indicates that when we have more "free" time, we'll be that much more productive in the time that we do work. The research also indicates that employment will be more evenly spread across the population. Let me stress this point: there will be *more work* available as those who have an over-abundance of work (and a deficit of time) choose to work less.

I could go on lauding the benefits of the kinds of shifts that economists like Schor, Victor and others propose. I could also continue to synthesize their research, arguments and conclusions. But my guess is that what I have proposed already either rings true or has completely alienated you in the sheer audacity of suggesting that we need an economic system that is not based on growth. If you fall into the latter category, I encourage you to take a moment and think about why you believe so strongly in the endless-growth economic model. As Victor points out: "Managing without growth seems like a very radical, even crazy idea, yet for all but the tiniest sliver of time since humans evolved, humanity has managed without growth" (2008a: 24). But can we really ever hope for a shift in policy from the use of economic referents like GDP to something that more broadly encompasses human happiness and environmental health (such as the Gross National Happiness index or the Happy Planet index)? Again, Victor reminds us that what we now take as common sense has only recently been learned. Only since the 1950s has economic growth become a policy priority. The growth model is the *aberration*! The fact that growth, along with so many other aspects of the business-as-usual market model, seems to be "logical" and "common sense" is by design: we have been told stories that celebrate this one way of understanding the world throughout our lives (and television tells these stories best). As Schor points out: "Neoliberal ideology has predisposed many to view market outcomes as natural or even fair, and has obscured the underlying biases, subsidies, and distortions associated with current market rules and structures" (2010: 161).

There is nothing natural, fair or patriotic about endlessly growing consumption. Yet as markets around the world sputter, "stimulus" and "growth" are the only stories we are told about what can be done. While the Earth continues to give us clear indications of extreme stress, new stories — our stories — are needed now more than ever before.

Many years ago we were warned. In 1975, E.F. Schumacher published *Small is Beautiful: Economics as if People Mattered*. His cautionary tale wisely warned that "if we squander our fossil fuels, we threaten civilization; but if we squander the capital represented by living nature around us, we threaten life itself" (17). At a time when the environmental movement was just beginning, his words were profound and affected many — and they continue to resonate today. In the conclusion to *Small is Beautiful*, Schumacher asks what he knows every reader is wondering: "What can I actually do?" He replies: "The answer is as simple as it is disconcerting: we can, each of us, work to put our own inner house in order" (297). Put our own inner house in order. Schumacher knew what so much research has now highlighted: the stories we are told extolling the virtues of materialism are lies. The vast majority of us are not happier, or wealthier (according to any definition of the word), because of our buying into these stories. So I propose that putting our own inner house in order involves, at a fundamental level, telling our own, utterly truthful, stories. Or as the authors of *Affluenza: The All-Consuming Epidemic* refer to it, "dreaming a new dream." One radical story we can tell is what "enough" would look like in our lives. Another radical story we can tell is what our lives look like when we are truly happy.

Tell Stories about Enough and Happiness

Once, years ago, I caught a glimpse of what it might look like if we lived in a culture that celebrated "enough." It was in a catalogue for the Canadian outdoor equipment manufacturer Mountain Equipment Co-op (MEC). MEC became a "consumer cooperative" (unlimited number of equal shares that cost $5) in 1971 and has since grown into a multi-million dollar a year business (as of 2011 there were over $260 million in annual sales and over three million members — the founding members having exactly the same $5 share as everyone else) with stores in sixteen locations across Canada (MEC website). While one can critique the materialistic role that such outdoor outfitters play in selling us the latest in new-age fabric fashions and other "must have" equipment for the outdoors, in one catalogue the focus of MEC's message was quite different. In that catalogue, early in the 2000s, the theme of "enough" was presented through testimonials highlighting that MEC's products, often purchased many years before, were so durable and continued to serve their purpose so well that no additional purchases were necessary. I was stunned because the concept was so foreign to me. Here was a company that (at least briefly — I haven't seen it done again, at least not with such a clear focus) was trying to make its customers as happy as possible by selling them as little as possible. The subtext was that less consumption would mean gentler treading on the earth.

I can only think of two other examples of messages of "enough," or at least "durability," as the central theme in a television advertisement (the fact that these examples are so dated should tell us how unusual these themes have been). The first message was started in 1967 by the appliance manufacturer Maytag, which began an advertising campaign developed by the Leo Burnett agency that was based on feeling sorry for the Maytag repair person. Why would we feel sorry? Because the Maytag product was so well made and durable that the poor repair person never had anything to do and was very lonely. Maytag's desire to make durability its claim to fame and propose that they were building a product that one might need to purchase only one of in one's lifetime, makes the ad campaign stand out.

The second message was in a Levi's ad that aired in the 1970s. I can't find many references to the ad online (no YouTube clip or mention of it on the Levi's website), but I remember the lyrics of the ad word-for-word (and the tune, although that's harder to convey here):

> I bought a pair of Levi's, they've really been around.
> I've taken them camping and laid them on the ground.
> They've been with me to parties and climbed up a tree.
> They've been to school so often they're nearly smart as me.
> And after years and years of wearing my Levi's in and out, I
>   began to notice that the knee wore out.
> So, I sewed on a patch… a flower here and there… and they look
>   so good again I can wear them anywhere.
> Now I'm wearing them as cutoffs and flying in the air.
> Don't you think it's time that I bought another pair?

Yes, the end result is that another pair of jeans is needed but only because the celebrated pair had been worn until they were falling apart. A single very durable pair of jeans. What a concept.

I propose that these are the kinds of stories we should be telling, sharing and celebrating with one another. What are your stories of clothing items you have owned forever that continue to serve you well? What are your stories of things you have recently given away or shared with others? What are your stories about what it's like to work in a garden and eat the results of your labour? What are your stories of celebrating your child's birthday in a creatively fun-filled and gift-free way? What are your stories of people who have inspired you to covet and consume less? And as you begin to explore what "enough" looks like in your life, also pay attention to how "enough" *feels*.

Let me tell you a beautiful story about enough. How does this story make me feel? Energized and optimistic. In 2006, a girl named Hannah Salwen was driving with her father when she made a powerful observation. As she looked out one side of her car there was a homeless person on the sidewalk and on the other side was someone driving a really expensive car. As a result of that moment of seeing inequality so clearly displayed for her, Hannah convinced her family to do something quite amazing: they downsized their seven-bedroom, 6,400 square foot home to a much cheaper house half the size. The family then donated $800,000 of the profits from the sale of the larger home to fund work in over a dozen villages in Ghana. In 2010, Hannah and her father wrote a book entitled *The Power of Half* to share the story of their experiences. The authors offer in the Introduction: "Yes, we're helping the world a bit. But in the process we are transforming our relationships with one another. And that has been the real surprise." One of the surprising ways that their lives were transformed? They discovered that in their new, smaller home, they spent more time together — and were happier (check out www.thepowerofhalf.com).

This is "enough," or Schor's notion of "plentitude," in action — when we embrace enough, we actually find ourselves amidst so much more than we had before. "The idea behind plentitude is that it moves us from a mix of incentives and imperatives that are no longer particularly efficient at delivering well-being (growth, work-and-spend, ecological degradation) to a way of living that a growing body of findings suggests will really make us better off" (2010: 176). After all, remember that there is a silver lining to all of the research that has been done illustrating that our material wealth has increased while our happiness has not. The counter-balancing research has explored how lower materialism is associated with a greater happiness and well-being. When you need to buy less stuff, your need to spend and earn decreases. As you need to spend less and earn less, your "time wealth" and "intrinsic values" (self-acceptance/personal growth, relatedness/intimacy, community feeling/helpfulness) can increase. When we are no longer on the work-and-spend treadmill, we have the time to pursue activities and relation-ships that make us fulfilled and happy. As this happens, share the stories of what the change looks like in your life and how the change feels. We tend to know what the "American Dream" is. This is a story we've all heard about the value of hard work in the name of material success. What about reframing that story? The non-profit organization Center for a New American Dream suggests that we move beyond pursuing "more" to pursuing "more of what matters — and less of what doesn't."

There is great power in the reclaiming of stories and storytelling. As smartMeme, an activist organization that uses storytelling as a tool for social change, reminds us: "We remember our lived experiences by converting them to narratives and integrating them into our personal and collective web of stories. Just as our bodies are made of blood and flesh, our identities are made of narratives" (Canning and Reinsborough 2009: 5). Indeed, happi-ness researcher Catherine O'Brien asks her students to think of the happiest person they know and share a story about that person. When I recently asked students how they would celebrate "enough," one group came up with the idea of circulating what they called "gratilogs": journals in which people could

share stories about what makes them grateful <http://gratitudelogproject.
wordpress.com>. We are so discouraged to think about our happiness and
gratitude, and the happiness and gratitude of those around us, that it may
be difficult to think of ourselves, friends and family in these ways, but when
we do, the quantity of material goods will probably not play a prominent
role in our stories.

Embrace Media Literacy

Sharing our stories will play an important role in the dismantling of materialism and endless-growth business as usual. We will of course also continue
to be told, everywhere and endlessly, mediated stories extolling the virtues
of materialism. While turning off our televisions and ceasing to expose
ourselves to its stories is a worthy goal, media literacy gives us the ability to
deconstruct and disempower those mediated stories that do reach us. Media
literacy — also known as media education, media awareness, media demys-
tification — is the proposition that we can learn to understand and think
critically about what we're watching on television (and what we're, more
generally, being exposed to in this ever-widening mediated communication
universe). Neil Postman offers at the end of *Amusing Ourselves to Death: Public
Discourse in the Age of Show Business*:

> It is the acknowledged task of the schools to assist the young in
> learning how to interpret the symbols of their culture. That this
> task should now require that they learn how to distance themselves
> from their forms of information is not so bizarre an enterprise that
> we cannot hope for its inclusion in the curriculum; even hope that
> it will be placed at the center of education. (1986: 163)

Indeed the establishment of "comprehensive media education programs
in schools" is one of the recommendations of the American Academy of
Pediatrics in their "Children, Adolescents, and Television" policy statement.

Media literacy scholar Barry Duncan tracks the origins of media literacy to the 1960s and the ideas of media philosopher Marshall McLuhan.
"Those ideas of looking at not just the content but the form of the media
was McLuhan's unique contribution" Duncan offers in a 2010 interview.
One of the first organizations dedicated to media literacy, the Association
for Media Literacy (AML), was started in 1978 in Canada. The AML's goals
include "helping students develop an informed and critical understanding of
the nature of the mass media, the techniques used by media industries, and
the impact of these techniques. [The AML] also aims to provide students with
the ability to create their own media products" (AML website). Media literacy,
in its various guises, has now spread across North America and around the

world. Its practitioners and advocates come from varied backgrounds and often approach the philosophy behind and practice of media literacy differently, but the fundamental importance of the "essence" of media literacy is increasingly embraced. Examples of advocates include the *Journal of Media Literacy Education*, the International Media Literacy Research Forum and the United Nations Educational, Scientific and Cultural Organization (UNESCO), which has as one of its missions to foster "information and media literate societies by encouraging the development of national information and media literacy policies, including in education" (UNESCO web site).

It seems that the biggest obstacle to media literacy as it continues into the twenty-first century is fighting for a widespread understanding of its importance. As Tessa Jolls, President and CEO of the Center for Media Literacy (CML) proposes:

> With stakes this big, and understanding of the new role of information and education in our society still so limited, it is imperative that media literacy become a movement of millions of people who seek to become excellent information managers, wise consumers, responsible producers and active participants in their communities. It is imperative that millions of people demand that these skills be formally taught to their children. (cited in Duncan 2010)

Indeed we should be raising our voices and demanding that media literacy become an integrated and integral part of school curricula everywhere. As part of this push for media literacy training, I suggest two key foci in the training: first, that students' creation of alternative stories be at the centre of media literacy; and second, that critical thinking about the media's — especially television's (on all of its screens) — promotion of materialism be at the centre of students' analyses and creation of alternative stories. I see this focus on producing alternative stories as being distinct from the insertion of messages into existing television stories (as the EMA does — discussed above). This activity should instead take a page from Adbusters Media Foundation's playbook and involve young people in creating both critiques of what they see on television and alternative visions of what they would like to see aired on television. These message production activities should also involve the sharing of the messages via social networking forums such as Facebook and YouTube.

The use of social networking to share students' critiques of television's role in promoting materialism is a great opportunity. This opportunity is highlighted both by young people's widespread use of social media — 75 percent of Americans aged thirteen to seventeen have a social network profile (Common Sense Media 2012) — and also by the role that such communication technology is playing in uprisings all over the world (e.g., the Arab

Spring, the Occupy Movement). That said, the use of social networks to share students' alternative stories should contain two caveats. First, media literacy must encourage a wariness of all screens. Students should be encouraged to be highly cognizant of the ways in which social networks also appeal to them as consumers. Facebook is a perfect example. As the most popular social networking site (according to checkfacebook.com there were over 877 million Facebook users mid-way through 2012), Facebook is constantly morphing into a slicker, more seamless, marketing machine. It couldn't be easier to sign up for a Facebook account, which helps you "connect and share with the people in your life." Facebook also helps advertisers and marketers connect and share with you and the people in your life. "People treat Facebook as an authentic part of their lives," the Facebook advertising section states, "so you can be sure you are connecting with real people with real interest in your products."

That's the thing about digital social networking — it may be decentralized and helpful in removing despotic heads of state and raising voices critical of inequity, but its role in further entrenching materialism is similarly decentralized (e.g., personalized) and profound. Facebook claims in its advertising section: "You can turn your advertising message into a trusted referral by including content from a user's friends who are already affiliated with your products." Therefore, media literacy must include discussions about the way in which Facebook and other social networking entities not only facilitate our connections for change but also facilitate entrenchment of the materialistic status quo.

The second caveat to students sharing stories via social networking is that the sharing should not stop there. Students should be encouraged to at least *attempt* to air their creations on television. The attempt will, no doubt, fail. As the activist group Adbusters Media Foundation discovered after years of trying, only a select few storytellers get to have their stories told on television (even when one is willing to pay for advertising time to tell the alternative story). When students encounter the reality that the public airwaves are only available to some, the next step is to involve them in an on-the-ground exercise that should involve lobbying for policy change to get more equitable access to television's storytelling.

Airing stories other than those of the corporate world on prime-time, mainstream commercial television, including stories that are highly critical of the corporate enterprise, will likely never happen. But having students demand that media policy-making entities like the Canadian Radio-television and Telecommunications Commission (CRTC) in Canada and the Federal Communications Commission (FCC) in the United States allow for a diversity of stories on television would be extremely valuable. Note that I don't mean that students should demand increased access to community or public television. They need to demand, at the very least, equal access to paid television advertising — any time, any station. Students need to understand the value of stories that explore our addiction to television, television's addiction to materialism, materialism's role in our endless-growth economy and the economy's endless assault on the environment. Students need to experience the difficulty of sharing such stories on mainstream media outlets.

A Finite Planet

When the Millennium Ecosystem Assessment (MEA) warned that "human activity is putting such a strain on the natural function of Earth that the ability of the planet's ecosystems to sustain future generations can no longer be taken for granted" (2005: 2), it also offered that with "appropriate actions it is possible to reverse the degradation of many ecosystem services over the next 50 years, but the changes in policy and practice required are substantial and not currently underway" (World Health Organization summary of MEA findings). The changes are not currently underway because the "human activity" being referenced isn't just that we drive more cars than ever before, or want bigger and bigger homes, or purchase new wardrobes each season, or eat more and more processed meat products, or rely increasingly on pharmaceuticals to make us feel better: it's that we are constantly being bombarded with stories telling us *that this is the only way life can look*. One of the places we should be able to find alternative stories about our relationship with the material world is environmental organizations. Yet alternative messages about materialism have been remarkably absent.

In the 1980s and 1990s, I volunteered and worked for several environmental organizations, including spending four and a half years as the information coordinator for Greenpeace Canada. In those years, and as far as I can tell in the years that followed, Greenpeace and most other North American environmental organizations have largely failed to explicitly address the issue of materialism. Certainly consumption is a topic that arises in campaigns. Environmental groups acknowledge the role of consumption in environmental degradation when they encourage people to avoid buying certain products. For example, over the years Greenpeace has encouraged people to avoid purchasing toxic materials such as polyvinylchloride (PVC) plastic and has had various "buycotts" (e.g., asking people to buy wood products from companies that log sustainably). Many environmental organizations encourage people to consume less energy. But a discourse that questions the stories we're told about and our fundamental relationship with the material world has been missing from the literature and campaigns of most environmental organizations. Indeed, Greenpeace, like so many environmental organizations, is very keen to sell us stuff — t-shirts, calendars, mugs, water bottles — as a way of showing support. The irony of this is highlighted by

Lisa Benton and John Short when they point out that you can "save the environment by buying products that you don't really need with sanitized nature images that can be conspicuously displayed in your already overly crowded cupboards and closets" (1999: 203).

We need environmental organizations to fundamentally challenge the materialism that sits at the heart of the environmental crisis. In order to do this, environmental organizations will have to occasionally disengage from the immediacy of the environmental crisis and help us envision an alternative future. Environmental organizations will have to speak directly to the Earth's devastation as a result of the endless-growth economic model. Environmental organizations will also have to act upon that alternative vision within their campaigns and operations. A good first step for a "Lessen the Earth's Load"-type campaign? No longer associate branded merchandise — t-shirts, calendars, mugs, stuffed animals — with one's environmental work, and explicitly encourage supporters to consume less.

One way, therefore, to tell new environmental stories is for environmental organizations to step up to the challenge of helping us tell stories about, and make the links between, materialism, the economy and the environment. Another way for us to tell new stories about the environment is to spend more time paying attention to the environment — *really* paying attention. Throughout the vast majority of human history, what we experienced and knew about not just the environment, but also our lives and those around us, came from our direct interaction with those around us and the world we share. Television has replaced interaction and made "normal" the world in which we currently live. What is increasingly apparent is that in order to tell new environmental stories, we'll have to get out there and experience the wonder of it for ourselves. And the wonder is everywhere. As Aristotle wisely offered us, "In all things of nature there is something of the marvelous" (*Parts of Animals* I.645a16) — go see for yourself!

Kill Your Television

Years ago I got a "Kill Your Television" bumper sticker. They were somewhat common for a while, and I still see them now and then. What I appreciate about the notion of "killing" our televisions is that by giving television animacy, or life (i.e., such that we would have to "kill" it), we also give it the respect it is due. And television is worthy of great respect. In just over fifty years, television has garnered a place of honour in most of our living rooms, and the number of sets in a home tends to outnumber the occupants. For many North Americans, the only activities that take up more time than watching television each day are sleeping and school or working. Interwoven within these realities is the real reason we should give television our respect: it tells stories that we *really* enjoy.

Nobody imposes television on us. It is a storyteller that we started enjoying when we were very young, and we continue to love its stories. We *choose* television. The problem is that we are paying a very high price for embracing television and welcoming it as our society's storyteller. The planet is reeling under the weight of television's celebration of endless-growth materialism.

In the preceding pages I have presented research and ideas encapsulated in the bumper sticker's message. There are the theories that help us make sense of how television affects us. There is the research that illustrates that materialism makes us unhappy. There are the statistics about the Earth's perilous state. There are the alternatives to endless-growth business as usual — alternative measures for assessing national well-being that include *our* well-being. In the end, the conclusions are straightforward; the bumper sticker is accurate. We are our stories. Television's stories are killing the planet. Kill your television. Tell your stories.

REFERENCES

Adamson, W. 1980. "Gramsci's Interpretation of Fascism." *Journal of the History of Ideas* 41(4): 615–633.

Adbusters. 1993. "Culture Jammer's Manifesto." December 25. <evolutionzone.com/kulturezone/futurec/culture.jammers.manifesto>.

Adbusters Media Foundation. <adbusters.org>.

Ader, C. 1995. "A Longitudinal Study of Agenda Setting for the Issue of Environmental Pollution." *Journalism and Mass Communication Quarterly* 7(2): 300–311.

Agrawala, S. 1998. "Context and Early Origins of the Intergovernmental Panel on Climate Change." *Climatic Change* 39: 605–620.

Alchon, S. 2003. *A Pest in the Land: New World Epidemics in a Global Perspective*. Albuquerque, NM: University of New Mexico Press.

American Academy of Pediatrics: Committee on Public Education. 2001. "Children, Adolescents and Television." *Pediatrics* 107(2): 423–426.

Anderson, C., L. Berkowitz, E. Donnerstein, R. Huesmann, J. Johnson, D. Linz, N. Malamuth, and E. Wartella. 2003. "The Influence of Media Violence on Youth." *Psychology Science in the Public Interest (American Psychological Society)* 4(30): 81–110.

Aristotle. "Parts of Animals. I.645a16." *Famous Quotes from 100 Great People*. Mobile Reference. <books.google.ca/books?id=cVAIBeCXxmwC&pg=PT99&lpg=PT99&dq=parts+of+animals+i.645a16&source=bl&ots=rTSSQx9Lna&sig=lvkRB1m_DcU2-sXx9hhP_Rxf00E&hl=en#v=onepage&q=parts%20of%20animals%20i.645a16&f=false>.

Asch, S. 1955. "Opinions and Social Pressure." *Scientific American* 193(5): 31–35.

Assadourian, E. 2010. "The Rise and Fall of Consumer Cultures." In *State of the World: Transforming Cultures from Consumerism to Sustainability*. Washington, DC: Worldwatch Institute.

Association for Media Literacy. <aml.ca/home>.

AutoShare. <autoshare.com>.

Ball-Rokeach, S., and M. DeFleur. 1976. "A Dependency Model of Mass Media Effects." *Communication Research* 3(1): 3–21.

Ball-Rokeach, S., M. Rokeach, and J. Grube. 1984. *The Great American Values Test*. New York: Free Press.

Bandura, A., D. Ross, and S.A. Ross. 1963. "Imitation of Film-Mediated Aggressive Models." *Journal of Abnormal and Social Psychology* 66(1): 3–11.

Barash, D. 2009. "We Are all Madoffs: Our Relationship to the Natural World Is a Ponzi Scheme." *The Chronicle of Higher Education*. August 31. <chronicle.com/article/We-Are-All-Madoffs/48182>.

Baron, R., and D. Kenny. 1986. "The Moderator-Mediator Variable Distinction in Social Psychological Research: Conceptual, Strategic, and Statistical Considerations." *Journal of Personality and Social Psychology* 51: 1173–1182.

Belk, R. 1985. "Trait Aspects of Living in the Material World." *Journal of Consumer Research* 12(3): 265–280.

Benton, L., and J. Short. 1999. *Environmental Discourse and Practice*. Malden, MA: Blackwell Publishers.

Berelson, B. 1949. "What 'Missing the Newspaper' Means." In P.F. Lazarsfeld and F.N. Stanton (eds.), *Communications Research 1948–1949*. New York, NY: Harper & Brothers.

Berman, M. 1989. *The Reenchantment of the World*. New York: Bantam Books.

Bill 64. 2008. "Cosmetic Pesticides Ban Act" <ontla.on.ca/web/bills/bills_detail.do?locale=en&BillID=1967>.

Bolls, P., D. Muehling, and K. Yoon. 2003. "The Effects of Television Commercial Pacing on Viewers' Attention and Memory." *Journal of Marketing Communications* 9(1): 17–28.

Bowling for Columbine. 2002. M. Moore (producer and director). United States: Alliance Atlantis.

Bowman, J., and K. Hanaford. 1977. "Mass Media and the Environment since Earth Day." *Journalism Quarterly* 54: 160–165.

Bowman, J., and T. Fuchs. 1981. "Environmental Coverage in the Mass Media: A Longitudinal Study." *International Journal of Environmental Studies* 18: 11–22.

Brand, J.E., and Greenberg, B.S. 1994. "Commercials in the Classroom: The Impact of Channel One Advertising." *Journal of Advertising Research* 34(1): 18–27.

Brink, J. 2008. *Imagining Head-Smashed-In: Aboriginal Buffalo Hunting on the Northern Plains*. Edmonton, AB: AU Press, Athabasca University.

Broad, W., and A. Revkin. 2003. "Has the Sea Given Up its Bounty?" *New York Times*. July 29. <nytimes.com/2003/07/29/science/has-the-sea-given-up-its-bounty.html?pagewanted=all&src=pm>.

Campaign for a Commercial-Free Childhood < commercialfreechildhood.org>.

Campbell, J. 1964. *The Masks of God, Vol. 3: Occidental Mythology*. New York, NY: Penguin (Non-Classics).

_____. 1949. *The Hero With a Thousand Faces*. New York, NY: Pantheon Books.

Campbell, J., with B. Moyer. 1991. *The Power of Myth*. <mythsdreamssymbols.com/functionsofmyth.html>.

Canadian Paediatric Society: Pscychosocial Paediatrics Committee. 2003. "Impact of Media Use on Children and Youth." *Paediatric Child Health* 8(5): 301–306. (Reaffirmed February 2011.)

Canning, D., and P. Reinsborough. 2009. "Re:Imaginging Change: An Introduction to Story-based Strategy." <smartmeme.org/downloads/smartMeme.ReImaginingChange.pdf>.

Cantril, H., H. Gaudet, and H, Herzog. 1940. *The Invasion from Mars: A Study in the Psychology of Panic*. Princeton, NJ: Princeton University Press.

Capitalism: A Love Story 2010. M. Moore (producer and diretor). United States: Alliance.

Carson, R. 1987 [1962]. *Silent Spring*. Boston, MA: Houghton Mifflin.

Centre for Media Literacy. <jmle.org/index.php/JMLE>.

Cheung, C., and C. Chan. 1996. "Television Viewing and Mean World Value in Hong Kong's Adolescents." *Social Behavior and Personality* 24: 351–364.

Churchill, G., and G. Moschis. 1979. "Television and Interpersonal Influences on Adolescent Consumer Learning." *Journal of Consumer Research* 6(June): 23–35.

Cohen, B. 1963. *The Press and Foreign Policy*. Princeton, NJ: Princeton University Press.

Cohen, P., and J. Cohen. 1996. *Life Values and Adolescent Mental Health*. Mahwah, NJ: Lawrence Erlbaum Associates.

Common Sense Media. 2012. "How Teens View Their Digital Lives." <http://www.commonsensemedia.org/sites/default/files/research/socialmediasociallife-final-061812.pdf>.

Consumer Electronics Association (CEA) as cited in "Facts and Figures on E-Waste Recycling." <electronicstakeback.com/wp-content/uploads/Facts_and_Figures>.

Consumerism. <merriam-webster.com/dictionary/consumerism>.

Cox, R. 2009. *Environmental Communication and the Public Sphere*. Thousand Oaks, CA: Sage Publications, Inc.

Cox, S. 2012. "Cooling a Warming Planet: A Global Air Conditioning Surge." <e360.yale. edu/feature/cooling_a_warming_planet_a_global_air_conditioning_surge/2550>.

———. 2010. *Losing Our Cool: Uncomfortable Truths About Our Air-Conditioned World (and Finding New Ways to Get Through the Summer)*. New York, NY: New Press.

Cresswell, J., and L. Thomas. 2009. "The Talented Mr. Madoff." January 25. <nytimes. com/2009/01/25/business/25bernie.html?pagewanted=all>.

Crowley, D., and P. Heyer. 2006. *Communication in History: Technology, Culture and Society* (fifth edition). New York: Allyn & Bacon.

Csikszentmihalyi, M. 2008. *Flow: The Psychology of Optimal Experience*. New York: Harper Perennial Modern Classics.

Dahlstrom, M., and D. Scheufele. 2010. "Diversity of Television Exposure and its Association with the Cultivation of Concern for Environmental Risks." *Environmental Communication* 4(1): 54–65.

Darling, D. *Encyclopedia of Science*. <daviddarling.info/encyclopedia/W/wave-particle_duality.html>.

Davies, S. 2004. "The Great Horse-Manure Crisis of 1894." *The Freeman: Ideas on Liberty* 54(7). <thefreemanonline.org/columns/our-economic-past-the-great-horse-manure-crisis-of-1894>.

DeGraaf, J., D. Wann, and T. Naylor. 2001. *Affluenza: The All-Consuming Epidemic*. San Francisco, CA: Berrett-Koehler Publishers.

Dietz, W., and S. Gortmaker. 1985. "Do We Fatten Our Children at the Television Set? Obesity and Television Viewing in Children and Adolescents." *Pediatrics* 75(5): 807–812.

Digital Detox Week. <adbusters.org/campaigns/digitaldetox>.

Discover Magazine. "25 Greatest Science Books of All Time." <discovermagazine. com/2006/dec/25-greatest-science-books/article_view?searchterm=top%20science%20books&b_start:int=1>.

Display Search. 2006. "Display Search Report Indicates Samsung Takes the Top Position in Global TV Units and Revenues" as cited in the Electronics Take Back Coalition "Facts and figures on e waste recycling." November 27. <tvtakeback.com/pdf/factnfigure.pdf>.

Duncan, B. 2010. "Voices of Media Literacy: International Pioneers Speak Out." <medialit.org/voices-media-literacy-international-pioneers-speak>.

Duncombe, S. 2007. *Dream: Re-Imagining Progressive Politics in an Age of Fantasy*. New York, NY: New Press.

Dunlap, R.E., K.D. VanLiere, A.G. Mertig, and R.E. Jones. 2000. "Measuring Endorsement of the New Ecological Paradigm: A Revised NEP Scale." *Journal of Social Issues* 56(3): 425–442.

Easterlin, R., and E. Crimmins. 1991. "Private Materialism, Personal Self-Fulfillment, Family Life, and Public Interest: The Nature, Effects, and Causes of Recent Changes in the Values of American Youth." *Public Opinion Quarterly* 55(4): 499–533.

Ehrlich P. 1968. *The Population Bomb*. New York: Ballantine Books.

Eisenstein, E. 1991. "The Rise of the Reading Public." In D. Crowley and P. Heyer (eds.), *Communication in History: Technology, Culture and Society*. New York, NY: Longman.

Ekins, Paul 1991. "The Sustainable Consumer Society: A Contradiction in Terms?" *International Environmental Affairs* 3(4): 243–258.

11th Hour. 2007. L. diCaprio (producer) and L. Conners (director). United States: Warner Independent Pictures.

Environmental Media Association. <ema-online.org>.

EPA (Environmental Protection Agency). 2005. As cited in "Facts and Figures on E-Waste Recycling." <electronicstakeback.com/wp-content/uploads/Facts_and_Figures>.

Epstein, S., R. Pacini, V. Dene-Raj, and H. Heier. 1996. "Individual Differences in Intuitive-Experiential and Analytical-Rational Thinking Styles." *Journal of Personality and Social Psychology* 71: 390–405.

eRideShare. <erideshare.com>.

Ewen, S. 1976. *Captains of Consciousness: Advertising and the Social Roots of the Consumer Culture.* New York: McGraw-Hill.

Facebook advertising examples. <facebook.com/advertising>.

Federal Trade Commission. 1998. Prepared statement of Deba A. Valentine, General Counsel for the U.S. Federal Trade Commission on "Pyramid Schemes." Presented at the International Monetary Fund's Seminar on Current Legal Issues Affecting Central Banks. Washington, DC, May 13. <http://www.ftc.gov/speeches/other/dvimfl6.shtm>.

Ferraro, R., and R. Avery. 2000. "Brand Appearances on Prime-Time Television." *Journal of Current Issues and Research in Advertising* 22(2): 1–15.

Fisch, S., R. Truglio, and C. Cole. 1999. "The Impact of Sesame Street on Preschool Children: A Review and Synthesis of 30 Years' Research." *Media Psychology* 1(2): 165–190.

Fisher, W. 1989. "Clarifying the Narrative Paradigm." *Communication Monographs* 56(1): 55–59.

___. 1985. "The Narrative Paradigm: In the Beginning." *Journal of Communication* 35(4): 74–89.

Food and Agriculture Organization of the United Nations. 2010. "Global Forest Resources Assessment." October 4. <fao.org/news/story/en/item/45904/icode>.

Freecycle. <freecycle.com>.

Frey, B., and A. Stutzer. 2002. *Happiness and Economics.* Princeton, NJ: Princeton University Press.

Gasquet, F. 1985. *The Great Pestilence AD 1348–1349: Now Commonly Known as the Black Death.* Whitefish, MT: Kessinger Publishing.

Gerbner, G., and L. Gross. 1976. "Living with Television: The Violence Profile." *Journal of Communication* 26(2): 173–199.

Gerbner, G., L. Gross, M. Morgan, and N. Signorielli. 1980. "The 'Mainstreaming' of America Violence Profile No. 11." *Journal of Communication* 30(30): 10–29.

Gilman, S., and X. Zhou. 2004. *Smoke: A Global History of Smoking.* London, UK: Reaktion Books.

Gitlin, T. 2003. *The Whole World Is Watching: Mass Media in the Making and Unmaking of the New Left.* Berkeley, CA: University of California Press.

Goldberg, M., and G. Gorn. 1978. "Some Unintended Consequences of TV Advertising to Children." *Journal of Consumer Research* 5: 22–29.

Gould, S.J. as cited by P. Forbes in "The Origin of Our Species by Chris Stringer — Review: A Valuable Guide to Human Prehistory." <guardian.co.uk/books/2011/jun/15/chris-stringer-origin-our-species-review>.

Good, J. 2009. "The Cultivation, Mainstreaming and Cognitive Processing of Environmentalists Watching Television." *Environmental Communication: The Journal of Nature and Culture* 3(3): 279–297.

____. 2007. "Shop 'til We Drop? Television, Materialism and Attitudes About the Natural Environment." *Mass Communication and Society* 10(3): 365–383.

____. 2007. Unpublished manuscript.

____. n.d. "An Open Letter to Loblaw's Galen Weston." <v1.theglobeandmail.com/servlet/story/RTGAM.20080629.wwebexclusive29/front/Front/frontBN/sympatico-front>.

Grant, T. 2010. "Excessive Personal Debt a Concern." *Globe and Mail.* October 25. <theglobeandmail.com/report-on-business/economy/excessive-personal-debt-a-concern/article1764981.>.

Gratilog Project. <http://gratitudelogproject.wordpress.com>.

Greenberg, B. 1974. "Gratifications of Television Viewing and Their Correlates for British Children." In J.G. Blumler and E. Katz (eds.), *The Uses of Mass Communication: Current Perspectives on Gratifications Research.* Beverly Hills, CA: Sage.

Greenpeace. n.d. "Forests – threats." <greenpeace.org/international/en/campaigns/forests/threats>.

Grontved, A., and F. Hu. 2011. "Television Viewing and Risk of Type 2 Diabetes, Cardiovascular Disease, and All-Cause Mortality: A Meta-Analysis." *Journal of the American Medical Association* 305(23): 2448–2455.

Gross Happiness Index. <grossnationalhappiness.com>.

Hall, S. 1997. "Representation & the Media." <mediaed.org/assets/products/409/transcript_409.pdf>.

Harmon, M. 2001. "Affluenza: Television Use and Cultivation of Materialism." *Mass Communication & Society* 4(4): 405–418.

Hayes, A. 2007. "Exploring the Forms of Self-Censorship: On the Spiral of Silence and the Use of Opinion Expression Avoidance Strategies." *Journal of Communication* 57: 785–802.

Herzog, H. 1944. "What Do We Really Know about Daytime Serial Listeners?" In P.F. Lazarsfeld and F.N. Stanton (eds.), *Radio Research 1942–1943.* New York: Duell, Sloan and Pearce.

Holbert, L., N. Kwak, and D. Shah. 2003. "Environmental Concern, Patterns of Television Viewing, and Pro-Environmental Behaviors: Integrating Models of Media Consumption and Effects." *Journal of Broadcasting and Electronic Media* 47(2): 177–196.

Howard-Williams, R. 2011. "Consumers, Crazies and Killer Whales: The Environment on New Zealand Television." *International Communication Gazette* 73(1–2): 27–43.

Huffman, K. 2009. "A Palmy Balm for the Financial Crisis." *The Sydney Morning Herald.* Febuary 9. <smh.com.au/news/opinion/a-palmy-balm-for-the-financial-crisis/2009/02/08/1234027847284.html>.

IECA (International Environmental Communication Association). <http://environmentalcomm.org>.

Inglehart, R. 2008. Changing Values Among Western Publics from 1970 to 2006. *West European Politics* 31(1–2): 130–146.

Intergovernmental Panel on Climate Change Group I: Policymakers Summary.1990. In J. Houghton, G. Jenkins and J. Ephraums (eds.), *Climate Change: The IPCC Scientific Assessment.* Cambridge, UK: Cambridge University Press.

____. Various dates. Reports. <http://www.ipcc.ch/publications_and_data/publications_and_data_reports.shtml#.UIgz-sXR5iY>.

International Media Literacy Research Forum. <imlrf.org>.

Jackson, T. 2009. *Prosperity Without Growth: Economics for a Finite Planet.* Washington, DC: Earthscan.

Jolls, T. 2011. "Voices of Media Literacy: International Pioneers Speak." <http://www.medialit.org/voices-media-literacy-international-pioneers-speak>. *Journal of Media Literacy Education*. <jmle.org/index.php/JMLE>.

Kasser, T. 2002. *The High Price of Materialism*. Cambridge, MA: MIT Press.

Kasser, T., and R. Ryan. 1996. "Further Examining the American Dream: Differential Correlates of Intrinsic and Extrinsic Goals." *Personality and Social Psychology* 65: 410–422.

Kasser, T., and K. Sheldon. 2000. "Of Wealth and Death: Materialism, Mortality Salience, and Consumption Behavior." *Psychological Science* 11(4): 348–351.

King, T. 2003. *The Truth About Stories: A Native Narrative*. Minneapolis, MN: University of Minnesota Press.

Klein, N. 2009. "Capitalism, Sarah Palin-Style." *The Progressive*. <progressive.org/klein0809.html>.

Kreshel, P. 1993. "Advertising Research in the Pre-Depression Years: A Cultural History." *Journal of Current Issues and Research in Advertising* 15(1): 59–75.

Kubey, R., and M. Csikszentmihalyi. 1990. *Television and the Quality of Life: How Viewing Shapes Everyday Experience*. Hillsdale, NJ: Lawrence Earlbaum Associates.

Lane, R. 2000. *The Loss of Happiness in Market Democracies*. New Haven, CT: Yale University Press.

Lasswell, H. 1948. "The Structure and Function of Communication in Society." In L. Bryson (ed.), *The Communication of Ideas*. New York: Harper & Row.

Leonard, A. n.d. "The Story of Stuff." <storyofstuff.org>.

Leonardo DiCaprio Foundation. <leonardodicaprio.org>.

Leroux, K. 1999. "Subliminal Messages: Primetime TV Programs Educate Viewers on the Environment." <jarodsafehouse.com/article_emagazine.php>.

Little, Amanda. 2003. "Julia Louis-Dreyfus and husband Brad Hall discuss their eco-friendly hideaway." *Grist*. <grist.org/article/griscom-house>.

Louv, R. 2005. *Last Child in the Woods: Saving our Children from Nature-Deficit Disorder*. Chapel Hill, NC: Algonquin Books.

Lovett, Richard. 2010. "Butchering Dinner 3.4 Million Years Ago." *Nature*. August 11.

Low, S., M. Chin, and M. Deurenberg-Yap. 2009. "Review on Epidemic of Obesity." *Annals Academy of Medicine* 38(1): 370–396.

Mander, J. 1991. *In the Absence of the Sacred: The Failure of Technology & the Survival of the Indian Nations*. San Francisco, CA: Sierra Club Books.

———. 1978. *Four Arguments for the Elimination of Television*. New York: Quill.

marino, dian. 1997. *Wild Garden: Art, Education, and the Culture of Resistance*. Toronto: Between the Lines.

Maslow, A. 1943. "A Theory of Human Motivation." *Psychological Review* 50: 370–396.

McChesney, R. 2008. *The Political Economy of Media: Enduring Issues, Emerging Dilemmas*. New York, NY: Monthly Review Press.

McChesney, R. <robertmcchesney.com>.

McComas, K., J. Shanahan, and J. Butler. 2001. "Environmental Content in Prime-Time Network and TV's Non-News Entertainment and Fictional Programs." *Society and Natural Resources* 14: 533–542.

McCombs, M., and A. Reynolds. 2002. "New Influence on Our Pictures of the World." In J. Bryant and D. Zillmann (eds.), *Media Effects* (second edition). Mahwah, NJ: Lawrence Erlbaum & Associates.

McCombs, M., and D. Shaw. 1972. "The Agenda-Setting Function of the Mass Media." *Public Opinion Quarterly* 36(2): 176–187.

McKibben, B. 2008. Interviewed by David Suzuki on the CBC radio program *The Current*. <cbc.ca/thecurrent/2008/10/october-10-2008.html>.

____. 1993. *The Age of Missing Information*. New York: Plume.

____. 1989. *The End of Nature*. New York: Anchor Books Doubleday.

McLuhan, M. 1964. *Understanding the Media: Extensions of Man*. New York: McGraw Hill.

McMahon, K. (director) and D. Sobelman (writer). 2002. *McLuhan's Wake* [video]. Available from National Film Board of Canada Library, 22-D Hollywood Avenue, Ho-Ho-Kus, New Jersey, 07423.

McQuail, D., J. Blumler, and J. Brown. 1972. "The Television Audience: A Revised Perspective." In D. McQuail (ed.), *Sociology of Mass Communication*. Middlesex, UK: Penguin.

MEC (Mountain Equipment Coop). <mec.ca>.

Merchant, C. 1989. *The Death of Nature: Women, Ecology and the Scientific Revolution*. San Francisco, CA: Harper & Row.

Millennium Ecosystem Assessment. 2005. *Living Beyond Our Means: Natural Assets and Human Well-being* [Statement from the Board]. Washington, DC: Author.

Morgan, M., and J. Shanahan. 1997. "Two Decades of Cultivation Research: An Appraisal and a Meta-Analysis." In B. Burelson (ed.), *Communication Yearbook 20*. Thousand Oaks, CA: Sage.

Museum of American Heritage (Technology and Development). n.d. "Let's Go to the Movies: The Mechanics of Moving Images." <moah.org/exhibits/archives/movies/technology_development.html>.

Museum of Science and Industry Collections Centre. n.d. "Telephone: First Patent in 1876 and First Commercial Exchange 1878. Early Manchester Telephone Exchanges." <mosi.org.uk/media/33871608/earlymanchestertelephoneexchanges.pdf >.

Myers, P., and F. Biocca. 1992. "The Elastic Body Image: The Effect of Television Advertising and Programming on Body Image Distortions in Young Women." *Journal of Communication* 42(3): 108–133.

National Communication Association's (NCA) history. <natcom.org/Default. aspx?id=108>.

National Review. n.d. "100 Best Non-fiction Books of the Century." <old.nationalreview.com/100best/100_books.html>.

National TV Turnoff week. <turnoffyourtv.com/turnoffweek/TV.turnoff.week.html>.

Naupa, A., and S. Lightner. 2005. *Histri Blong Yumi Long Vanuatu: An Educational Resource*. Vanuatu National Cultural Centre: Port Vila, Vanuatu.

Neighborgoods. <neighborgoods.com>.

Nelson, J. 1987. *The Perfect Machine: TV in the Nuclear Age*. Toronto: Between the Lines.

New Economics Foundation 2009. *The (Un)Happy Planet Index 2.0: Why Good Lives Don't Have to Cost the Earth*. <happyplanetindex.org/learn/download-report.html>.

____. 2006. *The (Un)Happy Planet Index: An Index of Human Wellbeing and Environmental Impact*. <happyplanetindex.org/learn/download-report.html>.

Nielsen. 2012. "Report: TV Continues to Hold the Lion's Share of Ad Dollars and Consumers' Media Time." <blog.nielsen.com/nielsenwire/online_mobile/report-tv-continues-to-hold-the-lion%E2%80%99s-share-of-ad-dollars-and-consumers-media-time/>.

____. 2010a. "State of the Media. TV Usage Trends Q2 2010." <blog.nielsen.com/nielsenwire/media_entertainment/state-of-the-media-tv-usage-trends-q2-2010>.

____. 2010b. "U.S. Homes Add Even More TV Sets in 2010." <blog.nielsen.com/nielsen-wire/consumer/u-s-homes-add-even-more-tv-sets-in-2010/#>.

____. 2009a. "Americans Watching More TV Than Ever; Web and Mobile Video Up too." May 20. <blog.nielsen.com/nielsenwire/online_mobile/americans-watching-more-tv-than-ever/>.

____. 2009b. "How Teens Use Media: A Nielsen Report on the Myths and Realities of Teen Media Trends." <blog.nielsen.com/nielsenwire/reports/nielsen_howteen-susemedia_june09.pdf>.

Noelle-Neumann, E. 1974. "The Spiral of Silence: A Theory of Public Opinion." *Journal of Communication* 24: 43–51.

Novic, K., and P. Sandman. 1974. "How Use of Mass Media Affects Views of Solutions to Environmental Problems." *Journalism Quarterly* 51: 448–452.

O'Brien, C. "Happiest Person You Know." <sustainablehappiness.ca>.

O'Guinn, T., and L.J. Shrum. 1997. "The Role of Television in the Construction of Consumer Reality." *Journal of Consumer Research* 23: 278–294.

O'Neil, D. 2011. "Early Modern Human Culture." <http://anthro.palomar.edu/homo2/mod_homo_5.htm>.

Olstad, S. 2009. "Cigarette Advertising." June 15. *Time*. <time.com/time/magazine/article/0,9171,1905530,00.html>.

Orange Bike projects. Overview of bike sharing at: <http://en.wikipedia.org/wiki/Bicycle_sharing_system>.

Parker-Pope, T. 2008. "A One-Eyed Invader in the Bedroom – Here's One Simple Way to Keep Your Children Healthy: Ban the Bedroom TV." *New York Times*. <nytimes.com/2008/03/04/health/04well.html>.

Patwardhan, P., and J. Yang. 2002. "Internet Dependency Relations and Online Consumer Behavior: A Media System Dependency Theory Perspective on Why People Shop, Chat, and Read News Online." *Journal of Interactive Advertising* 3(2): 57–69.

Pavlov, I. "Nobel Prize in Physiology or Medicine 1904." <nobelprize.org/nobel_prizes/medicine/laureates/1904/pavlov-bio.html>.

Peterson, R. 1991. "Physical Environment Television Advertisement Themes: 1979 and 1989." *Journal of Business Ethics* 10: 221–228.

Petty, R., and T. Cacioppo. 1986. *Communication and Persuasion: Central and Peripheral Routes to Attitude Change*. New York: Springer-Verlag.

PickupPal. <pickuppal.com>.

Pine, J. 2008. "Consumer Kids: How TV Advertisers Get into the Minds of Children." *Pediatrics for Parents* 25(7&8): 17–18.

Pipher, M. 1995. *Reviving Ophelia: Saving the Selves of Adolescent Girls*. New York: Ballantine Books.

Pleasant, A., J. Good, J. Shanahan, and B. Cohen. 2002. "The Literature of Environmental Communication." *Public Understanding of Science* 11(2): 197–205.

Postman, N. 1986. *Amusing Ourselves to Death: Public Discourse in the Age of Show Business*. New York: Penguin Books.

Putnam, R. 2001. *Bowling Alone: The Collapse and Revival of American Community*. New York: Touchstone.

____. 1995. "Bowling Alone: America's Declining Social Capital." *Journal of Democracy* 6(1): 65–78.

Raffaele, P. 2006. "In John They Trust: South Pacific villagers Worship a Mysterious American They Call John Frum — Believing He'll One Day Shower Their Remote Island with Riches." *Smithsonian Magazine* February. <smithsonianmag.com/people-places/john.html>.

Reeces Pieces example of product placement. <snopes.com/business/market/mandms.

asp>.

Reimer, B., and K.E. Rosengren. 1990. "Cultivated Viewers and Readers: A Lifestyle Perspective." In N. Signorielli and M. Morgan (eds.), *Cultivation Analysis: New Directions in Media Effects*. Newbury Park, CA: Sage.

Richins, M., and S. Dawson. 1992. "A Consumer Values Orientation for Materialism and its Measurement: Scale Development and Validation." *Journal of Consumer Research* 19: 303–316.

Rideout, V., and E. Hamel. 2006. "Information on Young Children: The Media Family: Electronic Media in the Lives of Infants, Toddlers, Preschoolers, and their Parents." <kff.org/entmedia/upload/7500.pdf>.

Roger and Me. 1989. M. Moore (producer and director). United States: Warner Bros. Pictures.

Rushdie, S. 1991. "Excerpts From Rushdie's Address: 1,000 Days 'Trapped Inside a Metaphor'." <nytimes.com/books/99/04/18/specials/rushdie-address.html>.

Saad, L. 2008. "U.S. Smoking Rate Still Coming Down: About One in Five American Adults Now Smoke." <gallup.com/poll/109048/us-smoking-rate-still-coming-down.aspx>.

Sagan, C. Cosmic Calendar. <visav.phys.uvic.ca/~babul/AstroCourses/P303/BB-slide.htm>.

Sale, J., L. Lohfeld, and K. Brazil. 2002. "Revisiting the Quantitative-Qualitative Debate: Implications for Mixed-Methods Research." *Quantity & Quality* 36: 43–53.

Salwen, K., and H. Salwen. 2010. *The Power of Half: One Family's Decision to Stop Taking and Start Giving Back.* Boston, MA: Houghton Mifflin Harcourt.

Scheufele, D., and P. Moy. 2000. "Twenty-Five Years of the Spiral of Silence: A Conceptual Review and Empirical Outlook." *International Journal of Public Opinion Research* 12(1): 3–28.

Schor, J. 2010. *Plentitude: The New Economics of True Wealth*. New York: Penguin Press.

———. 1992. *The Overworked American: The Unexpected Decline of Leisure*. New York: Basic Books.

Schumacher, E.F. 1975. *Small Is Beautiful: Economics as if People Mattered*. New York: Harper & Row.

Senecah, S. 2007. "Impetus, Mission, and Future of the Environmental Communication Commission Division: Are We Still On Track? Were We Ever?" *Environmental Communication: A Journal of Nature and Culture* 1(1): 21–33.

Shanahan, J., and K. McComas. 1999. *Nature Stories: Depictions of the Environment and Their Effects*. Cresskill, NJ: Hampton Press.

Shanahan, J., and M. Morgan. 1999. *Television and Its Viewers: Cultivation Theory and Research*. Cambridge, MA: Cambridge University Press.

Shayon, R. 1962. *The Eighth Art (23 Views of Television)*. Orlando, FL: Holt, Reinhart and Winston.

Shrum, L.J. 2007. "The Implications of Survey Method for Measuring Cultivation Effects." *Human Communication Research* 33: 64–80.

Shrum, L.J., J. Lee, J. Burroughs, A. Rindfleisch. 2011. "An Online Process Model of Second-Order Cultivation Effects: How Television Cultivates Materialism and Its Consequences for Life Satisfaction." *Human Communication Research* 37: 34–57.

———. 2005. "Television's Cultivation of Material Values." *Journal of Consumer Research* 32: 473–479.

Sirgy, M., L. Dong-Jin, R. Kosenko, L. Meadow, D. Rahtz, M. Cicic, X. Guang D. Yarsuvat, D. Blenkhorn, and N. Wright. 1998. "Does Television Viewership Play a Role in the Perception of Quality of Life?" *Journal of Advertising* 27(1): 125–142.

Smart, J. 1963. "Materialism." *The Journal of Philosophy* 60(22): 651–662.

Smithsonian Museum of Natural History. n.d. "What Does it Mean to be Human? Human Characteristics: Language and Symbols." <humanorigins.si.edu/human-characteristics/language>.

Smoking regulations. "Tobacco Timeline." <tobacco.org/resources/history/Tobacco_Historynotes.html>.

Stephens, M. n.d. "History of Television." *Grolier Encyclopedia* <nyu.edu/classes/stephens/History%20of%20Television%20page.htm>.

Terry O'Reilly <terryoreilly.ca>.

Television: First patent filed by Philo T. Farnsworth January 7, 1927 and issued on August 26, 1930. <google.com/patents?id=HRd5AAAAEBAJ&printsec=abstract&zoom=4#v=onepage&q&f=false>.

Thompson, J., and L. Heinberg. 1999. "The Media's Influence on Body Image Disturbance and Eating Disorders: We've Reviled Them, Now Can We Rehabilitate Them?" *Journal of Social Issues* 55(2): 339–353.

Trans Union and North American debt. <http://www.ctvnews.ca/canada/canadian-consumer-debt-level-hits-record-high-1.926424; http://www.abcnews4.com/story/19271837/transu>.

Trentmann, F. 2009. "Crossing Divides: Consumption and Globalization in History." *Journal of Consumer Culture* 9(2): 187–220.

U.S. Surgeon General's Scientific Advisory Committee on Television and Social Behavior. 1972. *Television and Growing Up: The Impact of Televised Violence* (DHEW Publication No. HSM 72-9086). Washington, DC.

Union of Concerned Scientists. 1994. "Statement of the Union of Concerned Scientists." <un.org/popin/icpd/conference/ngo/940909224555.html>.

UNESCO (United Nations Educational, Scientific and Cultural Organization). <unesco.org/new/en/unesco>.

United Nations Environment Programme. "The State of the Planet's Biodiversity." <unep.org/wed/2010/english/biodiversity.asp>.

University of Minnesota's mediated communication timeline. <mediahistory.umn.edu/timeline/>.

Victor, P. 2010. "Questioning Economic Growth." *Nature* 468: 370–371.

____. 2008a. *Managing Without Growth: Slower by Design, Not Disaster.* Northhampton, MA: Edward Elgar Publishing.

____. 2008b. Interview by David Suzuki on the CBC radio program *The Current.* <cbc.ca/thecurrent/2008/10/october-10-2008.html>.

Voice of America. 2010. "Family Sells Home for Hunger: Georgia Family Downsizes, Donates Half to Charity." February 15. <voanews.com/english/news/special-reports/american-life/Family-Sells-Home-for-Hunger-84299877.html>.

Wake, D., and V. Vredenburg. 2008. "Are We in the Midst of the Sixth Mass Extinction? A View from the World of Amphibians." *Proceedings of the National Academy of Sciences (PNAS)* 105(1): 11466–11473.

Wanta, W., and S. Ghanem. 2006. "Effects of Agenda Setting." In R. Preiss (ed.), *Mass Media Effects Research: Advances Through Meta-Analyses.* Mahwah, NJ: Lawrence Erlbaum Associates.

Wells, H.G. "War of the Worlds: Complete Transcript and Radio Broadcast of the 30 October 1938." <americanrhetoric.com/speeches/orsonwellswaroftheworlds.htm>.

Williams, R. 2005. *Culture and Materialism.* New York: Verso.

Wilson, J. 1998. *Talk and Log: Wilderness Politics in British Columbia.* Vancouver, BC: UBC Press.

Winn, M. 1977. *The Plug-In Drug: Television, Children, and the Family.* New York: Bantam Books.

World Health Organization. 2012. "Obesity and Overweight." <http://www.who.int/mediacentre/factsheets/fs311/en/index.html>.

___. 2011. "Got a Secure Job and Lots of Debt? Rejoice." *Globe and Mail* August 11. <theglobeandmail.com/globe-investor/personal-finance/got-a-secure-job-and-lots-of-debt-rejoice/article2125627>.

Wright, R. 2004. *A Short History of Progress.* Toronto: House of Anansi Press.

Yale, L., and M. Gilly. 1988. "Trends in Advertising Research: A Look at the Content of Marketing-Oriented Journals from 1976 to 1985." *Journal of Advertising* 17(1): 12–22.

Yellow Bike projects. Overview of bike sharing at: <http://en.wikipedia.org/wiki/Bicycle_sharing_system>.

Zimmerman, F., D. Christakis, and A. Meltzoff. 2004. "Early Television Exposure and Subsequent Attentional Problems in Children." *Journal of Pediatrics* 113(4): 708–713.

Zipcar. <http://www.zipcar.com>.

Zurawski, R. 2011. *Media Mediocrity Waging War Against Science: How the Television Makes Us Stoopid!* Halifax, NS: Fernwood Publishing.

INDEX